The FIDIC Contracts
Obligations of the Parties

Andy Hewitt
Construction Contracts and Claims Consultant
Hewitt Construction Consultancy

WILEY Blackwell

This edition first published 2014

© 2014 by John Wiley & Sons Ltd

Registered office: John Wiley & Sons, Ltd, The Atrium, Southern Gate, Chichester, West Sussex, PO19 8SQ, UK

Editorial offices: 9600 Garsington Road, Oxford, OX4 2DQ, UK

The Atrium, Southern Gate, Chichester, West Sussex, PO19 8SQ, UK

2121 State Avenue, Ames, Iowa 50014-8300, USA

For details of our global editorial offices, for customer services and for information about how to apply for permission to reuse the copyright material in this book please see our website at www.wiley.com/wiley-blackwell.

Library of Congress Cataloging-in-Publication Data

Hewitt, Andrew.
  The FIDIC contracts : obligations of the parties / Andy Hewitt.
    1 online resource.
  Description based on print version record and CIP data provided by publisher; resource not viewed.
  ISBN 978-1-118-29177-1 (Adobe PDF) – ISBN 978-1-118-29178-8 (ePub) – ISBN 978-1-118-29180-1 (pbk.) 1. Construction contracts. 2. Engineering contracts. 3. Architectural contracts. 4. Standardized terms of contract. I. Title.
  K891.B8
  343.07/8624–dc23
                                                          2013045169

A catalogue record for this book is available from the British Library.

Wiley also publishes its books in a variety of electronic formats. Some content that appears in print may not be available in electronic books.

Cover design by Workhaus

Set in 8/10 pt NewsGothicStd by Toppan Best-set Premedia Limited
Printed and bound in Malaysia by Vivar Printing Sdn Bhd

1 2014

Contents

About the Author

Andy Hewitt, FICCP, FCInstCES, FQSi, ACIOB, ACIArb, MDBF, is the principal of Hewitt Construction Consultancy, which specialises in construction contracts, claims and dispute resolution. He is also the developer and principal of 'Claims Class', which provides education and training on the subject of construction claims by way of distance learning and 2-day intensive training programmes.

Andy comes from a background of over 40 years experience in the construction industry, gained in the United Kingdom, Africa and the Middle East. He has worked on projects in the UK, Nigeria, Bahrain, Saudi Arabia, Jordan, Oman, Sudan, Tanzania and the United Arab Emirates. He has held senior contracts, commercial and project management positions with contractors, sub-contractors and consultants, including several years operating his own practice in the UK in the 1980s and '90s, which offered quantity surveying, estimating and project management services to contractors, subcontractors, consultants and private clients. During his career, he has been involved in a wide variety of construction projects including super high-rise, large low-rise housing development, hotels, shopping malls, airports, hospitals, heavy civil engineering, process plants, marine works, water treatment and reclamation plants, pipelines, marine works, desalination plants and royal palaces.

His project and company management skills together with experience gained in a variety of disciplines and projects within the construction industry have led him, in the latter part of his career, to specialise in the field of construction contractual issues, claims and dispute management.

Andy's first book, *Construction Claims & Responses: effective writing and presentation*, has enjoyed international success and was selected as book of the month by Construction Books Direct during its first month of publication. Details of his consultancy practice may be found at http://www.hewittconstructionconsultancy.com. Information on courses offering training and education in construction claims may be found at http://constructionclaimsclass.com.

Foreword

Andy Hewitt, having written his successful first book, *Construction Claims and Responses*, embarked upon and completed a second. His first book passed on to its readers his vast knowledge and experience of writing and responding to claims. The second book demonstrates that, as a result of working in the construction industry for all his working life, Andy has recognised that many of those engaged regularly on projects either have a reluctance to examine the fine print set out in the Conditions of Contract or, having read them experience difficulty understanding what is required to enforce the rights and obligations of the parties. This is hardly surprising: for example the Conditions of Contract for Construction for Building and Engineering Works Designed by the Employer (Red Book), which is the most widely used of the FIDIC forms, comprises 60 pages of close printed conditions and another 30 pages of guide notes and examples.

To overcome this problem Andy has produced a book which sets out the rights and obligations of the parties in a tabular format.

The first thing that strikes the reader is the manner in which the information is contained in the book, which makes it easy to follow and understand.

The book sets out separately the rights and obligations of the Employer Contractor and Engineer and covers every clause in the differing FIDIC forms including the Consultant's Model Service Agreement and the workings of the Adjudication Board.

Those who use the FIDIC forms of contract quickly discover that a failure to follow the procedures as set out in the Conditions of Contract can have serious financial consequences for the Employer, Contractor and Subcontractor. For example, if a contractor employed using the Red Book encounters unforeseen adverse ground conditions, there is a procedure to be followed to ensure that adequate financial compensation is secured. The book provides in a succinct manner the four essential procedural steps that the Contractor must take to ensure proper financial reimbursement. The Contractor's entitlements to additional time for completion and for additional payment resulting from a significant number of events that may occur on any site are catered for in the contract. The book provides a six step guide to ensure the necessary procedures are followed. This is the format for the whole of the book.

It is clear that there is less chance of failure to observe contract compliance using this book than from reliance on reading through the appropriate clauses in the contract.

A big plus is that those using the book will find answers to queries relating to contractual issues arising from the FIDIC contracts' conditions in a fraction of the time it would take if it were necessary to study the full text.

For those using the FIDIC forms for the first time, or infrequently, this book is a must, whilst experienced users will find it a valuable memory jogger. Whichever category the reader falls into, using this book should improve performance.

The book is ideal for engineers, quantity surveyors, contract managers and any person whose job it is to understand the workings of a FIDIC contract.

Roger Knowles

Acknowledgements

I would like to acknowledge the International Federation of Consulting Engineers who has kindly given me permission to reproduce sections from the FIDIC suite of contracts. FIDIC's contact details are as follows:

World Trade Centre II
PO Box 311
Geneva
Switzerland
Telephone: +41 22 799 4905
Fax: −41 22 799 4901
Email: fidic@fidic.org
www: www.fidic.org

Introduction

FIDIC

FIDIC is the International Federation of Consulting Engineers and its members are comprised of national associations of consulting engineers. Today, membership covers 94 countries of the World and since it's inauguration in 1913, the FIDIC standard forms of contract have become the international standard for contracts of an international nature and are widely used in developing countries that have not produced their own standard forms of contract.

In 1999 FIDIC published a 'suite' of contracts, which became known as 'the Rainbow Suite' because of the various colours of the covers, these have been refined and developed in the intervening years and today the suite comprises the following contracts:

Conditions of Contract for Construction for Building and Engineering Works Designed by the Employer (the Red Book).

Conditions of Contract for Construction, MDB[1] Harmonised Edition for Building and Engineering Works Designed by the Employer (the Harmonised Red Book).

Conditions of Subcontract for Construction for Building and Engineering Works Designed by the Employer (the Red Book Subcontract).

Conditions of Contract for Plant and Design-Build for Electrical and Mechanical Plant, and for Building and Engineering Works, Designed by the Contractor (the Yellow Book).

Conditions of Contract for EPC[2] / Turnkey Projects (the Silver Book).

Conditions of Contract for Design, Build and Operate Projects (the Gold Book).

[1] MDB – Multilateral Development Bank

[2] EPC – Engineer, Procure and Construct

Short Form of Contract (the Green Book).

Conditions of Contract for Dredging and Reclamation Works (the Blue-Green Book).

Client/Consultant Model Services Agreement (the White Book).

The Purpose of this Book

Any document containing conditions of contract should essentially set out the rights and obligations of the parties, but if we look at a comprehensive set of conditions such as those contained in the FIDIC suite, it may be seen that it becomes necessary to set out rules, procedures and requirements and in some cases to provide additional clarification of various provisions. This is because of the complexity of the work that is the subject of the contact and this inevitably results in a lengthy and complicated document.

One of the principles of a contract is that the parties should work positively to help each other to perform the contract and this is usually set out in the conditions by describing the obligations of each party. In most cases, obligations are made specific by including the word 'shall', for example; *'The Employer* shall *give the Contractor right of access to, and possession of, all parts of the Site . . .'*. In other cases however, the obligations are implied, for example, a provision that states that *'The Contractor shall submit a Statement in six copies to the Engineer after the end of each month, in a form approved by the Engineer . . .'*, places an obligation on the Engineer to work positively to approve the form of the statement.

The rights of the parties do not always include obligations. For example, *'If the Tests on Completion are being unduly delayed by the Contractor, the Engineer* may, *by notice require the Contractor to carry out the Tests within 21 days after receiving the notice'*. This gives the Engineer the right to require the Contractor to carry out the tests, but it is not an obligation. If the Engineer does decide to exercise this right however, he is obliged to inform the contractor by way of a notice.

Inevitably, in a long and complicated document, important information can become difficult to find or not be immediately apparent. This is particularly so for persons who are not 'contractually minded' or are not familiar with the particular contract conditions being used and the consequences of a party not being aware of its own obligations could easily lead to a breach of contract with possible costly consequences. It is also important to be aware of the other party's obligations to ensure that they perform them properly and if they do not do so, be aware of any rights that may provide compensation under such circumstances.

The purpose of this book is therefore to set out the obligations of each of the parties for each of the conditions of contract contained in the FIDIC suite in a simple and easily understood manner and this has been done by listing each party's obligations separately in a table format.

The conditions often require the parties to perform their obligations within specific times and the tables therefore also include details of any time frames that are included within the conditions.

Finally, any specific consequences of non-performance of the obligations are listed. In this regard, it should be noted that only the consequences contained in the conditions are included. If none are listed, this does not mean that there are no consequences and in fact any failure to perform an obligation would be a breach of contract. The fact that no specific remedy is listed just means that the law rather than the contact conditions would determine the remedy for any breach.

Chapter 1

The Red Book

Conditions of Contract for Construction, for Building and Engineering Works Designed by the Employer, First Edition 1999

The FIDIC Contracts: Obligations of the Parties, First Edition. Andy Hewitt.
© 2014 John Wiley & Sons, Ltd. Published 2014 by John Wiley & Sons, Ltd.

THE OBLIGATIONS OF THE EMPLOYER

CLAUSE	OBLIGATIONS	TIME FRAME	SPECIFIC CONSEQUENCES OF NON-COMPLIANCE
	GENERAL CONDITIONS		
1 General Provisions			
1.6 Contract Agreement	Enter into a Contract Agreement with the Contractor.	Within 28 days after the Contractor receives the Letter of Acceptance, unless agreed otherwise.	None.
1.8 Care and Supply of Documents	a) Keep the Specification and Drawings in custody and care. b) Supply two copies of the Contract and of each subsequent Drawing to the Contractor. c) Give notice of errors or defects in any document prepared by the Contractor for use in executing the Works.	a) None. b) None. c) Promptly.	None.
1.13 Compliance with Laws	Obtain the planning, zoning or similar permission for the Permanent Works, and any other permissions described in the Specification as having been (or being) obtained by the Employer.	None.	None.
2 The Employer			
2.1 Right of Access to the Site	a) Give the Contractor right of access to, and possession of, all parts of the Site. b) Give the Contractor possession of any foundation, structure, plant or means of access if required.	a) Within the time (or times) stated in the Appendix to Tender, or if not stated, to enable the Contractor to proceed in accordance with the programme submitted under Sub-Clause 8.3 *[Programme]*. b) In the time and manner stated in the Specification.	Contractor shall be entitled to an extension of time and payment of Cost plus reasonable profit.

THE OBLIGATIONS OF THE EMPLOYER (continued)

CLAUSE	OBLIGATIONS	TIME FRAME	SPECIFIC CONSEQUENCES OF NON-COMPLIANCE
2.2 Permits, Licences or Approvals	Provide reasonable assistance to the Contractor at the request of the Contractor: a) By obtaining copies of the Laws of the Country which are relevant but not readily available, and b) For the Contractor's applications for any permits, licences or approvals required by the Laws of the Country.	None.	None.
2.3 Employer's Personnel	Ensure that Employer's personnel and Employer's other contractors cooperate with the Contractor and take actions similar to those which the Contractor is required to take under Sub-Clause 4.8 [Safety Procedures] and under Sub-Clause 4.18 [Protection of the Environment].	None.	None.
2.4 Employer's Financial Arrangements	Submit reasonable evidence that financial arrangements have been made and are being maintained which will enable the Employer to pay the Contract Price.	Within 28 days after receiving any request from the Contractor.	Contractor entitled to suspend work, reduce the rate of work and to an extension of time and additional payment as a result of such actions (Sub-Clause 16.1).
2.5 Employer's Claims	Give notice and particulars to the Contractor if the Employer considers himself to be entitled to any payment under any Clause of the Conditions or otherwise in connection with the Contract and/or to any extension of the Defects Notification Period (this obligation may be carried out by the Engineer).	As soon as practicable after becoming aware of the event or circumstances giving rise to the claim.	None.

CLAUSE	OBLIGATIONS	TIME FRAME	SPECIFIC CONSEQUENCES OF NON-COMPLIANCE
3 The Engineer			
3.1 Engineer's Duties and Authority	a) Appoint the Engineer to carry out the duties assigned to him in the Contract. b) Not to impose further constraints on the Engineer's authority, except as agreed with the Contractor.	None.	None.
3.4 Replacement of the Engineer	a) Give notice to the Contractor of the name, address and relevant experience of the intended replacement Engineer. b) Not to replace the Engineer with a person against whom the Contractor raises reasonable objection by notice to the Employer with supporting particulars.	a) Not less than 42 days before the intended date of replacement. b) None.	None.
4 The Contractor			
4.2 Performance Security	a) Cooperate with the Contractor to agree the entity, country (or other jurisdiction) for the issue of the Performance Security. b) Cooperate with the Contractor to agree the form of Performance Security if not in the form annexed to the Particular Conditions. c) Not to make a claim under the Performance Security, except for amounts to which the Employer is entitled under the Contract (as listed). d) Return the Performance Security to the Contractor.	a) None. b) None. c) None. d) Within 21 days after receiving a copy of the Performance Certificate.	None.
4.10 Site Data	a) Make available to the Contractor all relevant data in the Employer's possession on sub-surface and hydrological conditions at the Site, including environmental aspects. b) Make available to the Contractor all such data which come into the Employer's possession after the Base Date.	a) Prior to the Base Date. b) None.	None.

THE OBLIGATIONS OF THE EMPLOYER

Red Book

THE OBLIGATIONS OF THE EMPLOYER (continued)

CLAUSE	OBLIGATIONS	TIME FRAME	SPECIFIC CONSEQUENCES OF NON-COMPLIANCE
4.20 Employer's Equipment and Free-Issue Material	a) Make the Employer's Equipment (if any) available for the use of the Contractor in the execution of the Works in accordance with the details, arrangements and prices stated in the Specification. b) Supply, free of charge, the "free-issue materials" (if any) in accordance with the details stated in the Specification. c) Rectify the notified shortage, defect or default in the free-issue materials.	a) As specified in the Contract. b) As specified in the Contract. c) Immediately.	None.
4.24 Fossils	Take possession and care of fossils, coins, articles of value or antiquity and structures and other remains, or items of geological or archaeological interest found on the Site.	None.	None.
10 Employer's Taking Over			
10.1 Taking Over of the Works and Sections	Take over the Works.	When completed in accordance with the Contract.	None.
10.2 Taking Over of Parts of the Works	Not to use any part of the Works (other than as a temporary measure which is either specified in the Contract or agreed by both Parties) unless and until the Engineer has issued a Taking-Over Certificate for this part.	None.	If the Contractor incurs Cost as a result of the Employer taking over and/or using a part of the Works, the Contractor shall be entitled to payment of any such Cost plus reasonable profit.

CLAUSE	OBLIGATIONS	TIME FRAME	SPECIFIC CONSEQUENCES OF NON-COMPLIANCE
11 Defects Liability			
11.2 Cost of Remedying Defects	Notify the Contractor (or ensure that notice is given by others) of any work to be remedied if due to any cause outside the provisions of the Contract.	Promptly.	None.
11.4 Failure to Remedy Defects	In the case of failure by the Contractor to remedy any defect or damage, notify the Contractor (or ensure that notice is given by others) of the date by which the defect or damage is to be remedied.	Within reasonable time.	None.
14 Contract Price and Payment			
14.2 Advance Payment	a) Make the advance payment, as an interest-free loan for mobilisation. b) Cooperate with the Contractor to approve the form of the advance payment guarantee.	a) As stated in the Appendix to Tender and when the Contractor submits a guarantee in accordance with this Sub-Clause. b) None.	a) Contractor entitled to suspend work, reduce the rate of work and to an extension of time and additional payment as a result of such actions (Sub-Clause 16.1) and payment of financing charges to the Contractor (Sub-Clause 14.8). b) None.
14.5 Plant and Material intended for the Works	Cooperate with the Contractor to approve the form of a bank guarantee for shipped Plant and Materials.	None.	None.

THE OBLIGATIONS OF THE EMPLOYER

Red Book

THE OBLIGATIONS OF THE EMPLOYER (continued)

CLAUSE	OBLIGATIONS	TIME FRAME	SPECIFIC CONSEQUENCES OF NON-COMPLIANCE
14.7 Payment	a) Pay the first instalment of the advance payment. b) Pay the amount certified in each Interim Payment Certificate. c) Pay the amount certified in the Final Payment Certificate.	a) Within 42 days after issuing the Letter of Acceptance or within 21 days after receiving the documents in accordance with Sub-Clause 4.2 *[Performance Security]* and Sub-Clause 14.2 *[Advance Payment]*, whichever is later. b) Within 56 days after the Engineer receives the Statement and supporting documents. c) Within 56 days after the Employer receives the Final Payment Certificate.	Contractor entitled to suspend work, reduce the rate of work and to an extension of time and additional payment as a result of such actions (Sub-Clause 16.1) and payment of financing charges to the Contractor (Sub-Clause 14.8).
14.9 Payment of Retention Money	a) Pay the Contractor the first half of the Retention Money. b) Pay the Contractor the outstanding balance of the Retention Money.	a) When a Taking-Over Certificate has been issued and when the Engineer has certified payment. b) Promptly after the latest of the expiry of the Defects Notification Periods and when the Engineer has certified payment.	The Contractor is entitled to receive financing charges (Sub-Clause 14.7).
14.15 Currencies of Payment	Pay the Contractor in the currency or currencies named in the Appendix to Tender.	None.	None.
15 Termination by Employer			
15.2 Termination by Employer	a) Give notice of intention to terminate the Contract. b) Give notice of release of the Contractor's Equipment and Temporary Works.	a) 14 days prior to the termination date, or immediately in the case of the Contractor becoming bankrupt or gives or offers bribes or gratuities (or similar as defined in the clause). b) On completion of the Works.	None.

CLAUSE	OBLIGATIONS	TIME FRAME	SPECIFIC CONSEQUENCES OF NON-COMPLIANCE
15.4 Payment after Termination	Pay the balance due to the Contractor after recovering any losses, damages and extra costs.	None.	None.
15.5 Employer's Entitlement to Termination	a) Give notice of intention to terminate the Contract. b) Return the Performance Security. c) Not to terminate the Contract in order to execute the Works himself or to arrange for the Works to be executed by another contractor.	a) 28 days prior to the termination date. b) None. c) None.	None.
16 Suspension and Termination by Contractor			
16.4 Payment on Termination	a) Return the Performance Security to the Contractor. b) Pay the Contractor in accordance with Sub-Clause 19.6 *[Optional Termination, Payment and Release]*. c) Pay to the Contractor the amount of any loss of profit or other loss or damage sustained by the Contractor as a result of this termination.	Promptly.	None.
17 Risk and Responsibility			
17.1 Indemnities	Indemnify and hold harmless the Contractor, the Contractor's Personnel and their respective agents, against and from all claims, damages, losses and expenses in respect of bodily injury, disease or death which is attributable to any negligence, wilful act or breach of the Contract by the Employer, the Employer's Personnel or agents.	None.	None.

THE OBLIGATIONS OF THE EMPLOYER

Red Book

THE OBLIGATIONS OF THE EMPLOYER (continued)

CLAUSE	OBLIGATIONS	TIME FRAME	SPECIFIC CONSEQUENCES OF NON-COMPLIANCE
17.5 Intellectual and Industrial Property Rights	a) Give notice of any claim under this clause. b) Indemnify and hold the Contractor harmless against and from any claim, which is or was an unavoidable result of the Contractor's compliance with the Contract or as a result of any Works being used by the Employer. c) If requested by the Contractor, assist in contesting the claim. d) Not to make any admission, which might be prejudicial to the Contractor.	a) Within 28 days of receiving a claim. b) None. c) None. d) None.	a) Waiver of right to indemnity. b) None. c) None. d) None.
18 Insurance			
18.1 General Requirements for Insurances	a) Cooperate with the Contractor to approve the terms of insurances. b) Effect and maintain the insurances in terms consistent with the details annexed to the Particular Conditions, wherever the Employer is the insuring Party. c) Submit evidence that the insurance has been effected, provide copies of the policies and submit evidence of payment. d) Inform the insurers of any relevant changes to the execution of the Works and ensure that insurance is maintained. e) Not to make any material alteration to the terms of any insurance without approval of the Contractor.	a) None. b) None. c) Within the time frames stipulated in the Contract Data. d) As appropriate. e) None.	a) None. b) None. c) The Contractor may effect the insurance and the Contract Price shall be adjusted. d) None. e) None.

CLAUSE	OBLIGATIONS	TIME FRAME	SPECIFIC CONSEQUENCES OF NON-COMPLIANCE
19 Force Majeure			
19.2 Notice of Force Majeure	Give notice to the Contractor in the case that the Employer is, or will be prevented from performing the Employer's obligations by Force Majeure.	Within 14 days of becoming aware of the event.	None.
19.3 Duty to Minimise Delay	a) Use all reasonable endeavours to minimise any delay in the performance of the Contract as a result of Force Majeure. b) Give notice when the effects of the Force Majeure cease.	None.	None.
20 Claims, Disputes and Arbitration			
20.2 Appointment of the Dispute Adjudication Board	a) Jointly appoint the Dispute Adjudication Board (DAB). b) Not to consult the DAB without the agreement of the Contractor. c) Not to act alone in the termination of any member of the DAB.	a) By the date stated in the Appendix to Tender. b) None. c) None.	a) The appointing entity or official named in the Appendix to Tender shall appoint (Sub-Clause 20.3). b) None. c) None.
20.4 Obtaining Dispute Adjudication Board's Decision	a) Make available to the DAB additional information, access to the Site, and appropriate facilities as the DAB may require. b) Give effect to a DAB decision unless and until it is revised in an amicable settlement or an arbitral award.	a) Promptly. b) None.	a) None. b) The matter may be referred to arbitration (Sub-Clause 20.7).
20.5 Amicable Settlement	In the case of a notice of dissatisfaction being issued, attempt to settle the dispute amicably.	Within 56 days of the notice.	Arbitration may be commenced.

THE OBLIGATIONS OF THE EMPLOYER

Red Book

THE OBLIGATIONS OF THE EMPLOYER (continued)

CLAUSE	OBLIGATIONS	TIME FRAME	SPECIFIC CONSEQUENCES OF NON-COMPLIANCE
GENERAL CONDITIONS OF DISPUTE ADJUDICATION AGREEMENT			
2 General Provisions	Give notice to the Dispute Adjudication Board Member that the Dispute Adjudication Agreement has taken effect.	On the Commencement Date, or upon all parties signing the Dispute Adjudication Agreement whichever is the later.	None.
5 General Obligations of the Employer and the Contractor	a) Not to request advice from, or consultation with a DAB Member regarding the Contract otherwise than in the normal course of the DAB's activities. b) In the case of the DAB Member being required to make a site visit or attend a hearing, provide appropriate security for a sum equivalent to the reasonable expenses to be incurred by the Member (this may be undertaken by the Contractor).	None.	None.
6 Payment	a) Pay one half of the DAB fees to the Contractor. b) In the Case of the Contractor failing to pay the DAB Member, pay the due fees.	a) Within the monthly payments to the Contractor. b) None.	The DAB Member may suspend services or resign the appointment.
Annex – Procedural Rules			
2.	Jointly agree the timing of and agenda for each site visit by the DAB.	None.	Timing and agenda shall be decided by the DAB.

CLAUSE	OBLIGATIONS	TIME FRAME	SPECIFIC CONSEQUENCES OF NON-COMPLIANCE
3.	a) Attend site visits by the DAB. b) Co-ordinate site visits by the DAB in co-operation with the Contractor. c) Ensure the provision of appropriate conference facilities and secretarial and copying services to the DAB.	None.	None.
4.	a) Furnish to the DAB one copy of all documents, which the DAB may request. b) Copy the Contractor on all communications between the DAB and the Employer.	None.	None.

THE OBLIGATIONS OF THE EMPLOYER

THE OBLIGATIONS OF THE CONTRACTOR

CLAUSE	OBLIGATIONS	TIME FRAME	SPECIFIC CONSEQUENCES OF NON-COMPLIANCE
	GENERAL CONDITIONS		
1 Definitions			
1.6 Contract Agreement	Enter into a Contract Agreement.	Within 28 days after the Contractor receives the Letter of Acceptance, unless agreed otherwise.	None.
1.8 Care and Supply of Documents	a) Supply to the Engineer six copies of each of the Contractor's Documents. b) Keep, on the Site, a copy of the Contract, publications named in the Specification, the Contractor's Documents (if any), the Drawings and Variations and other communications given under the Contract. c) In the case of an error in a document, give notice to the Employer.	a) None. b) None. c) Promptly.	None.
1.9 Delayed Drawings or Instructions	a) Give notice to the Engineer whenever the Works are likely to be delayed or disrupted, if any necessary drawing or instruction is not issued to the Contractor within a particular time. b) Give a further notice to the Engineer if the Contractor suffers delay and/or incurs Cost.	a) Within a reasonable time. b) As soon as practicable and not later than 28 days after the Contractor became aware, or should have become aware of the event or circumstance (Sub-Clause 20.1).	a) None. b) Loss of entitlement to an extension to the Time for Completion and additional payment (Sub-Clause 20.1).
1.12 Confidential Details	Disclose all such confidential and other information as the Engineer may reasonably require in order to verify the Contractor's compliance with the Contract.	None.	None.

CLAUSE	OBLIGATIONS	TIME FRAME	SPECIFIC CONSEQUENCES OF NON-COMPLIANCE
1.13 Compliance with Laws	a) Comply with applicable Laws. b) Give all notices, pay all taxes, duties and fees, and obtain all permits, licences and approvals, as required by the Laws.	None.	None.
1.14 Joint and Several Liability	a) In the case of a joint venture, consortium or other unincorporated grouping of two or more persons, notify the Employer of the leader. b) Not to alter the composition or legal status of the joint venture without the prior consent of the Employer.	None.	None.
2 The Employer			
2.1 Right of Access to the Site	Give notice to the Engineer if the Contractor suffers delay and/or incurs Cost as a result of failure to give right of access to, and possession of the Site.	As soon as practicable and not later than 28 days after the Contractor became aware, or should have become aware of the event or circumstance (Sub-Clause 20.1).	Loss of entitlement to an extension to the Time for Completion and additional payment (Sub-Clause 20.1).
3 The Engineer			
3.3 Instructions of the Engineer	a) Only take instructions from the Engineer, or from an assistant to whom the appropriate authority has been delegated. b) Comply with the instructions given by the Engineer or delegated assistant on any matter related to the Contract.	None.	None.

THE OBLIGATIONS OF THE CONTRACTOR (continued)

CLAUSE	OBLIGATIONS	TIME FRAME	SPECIFIC CONSEQUENCES OF NON-COMPLIANCE
4 The Contractor			
4.1 Contractor's General Obligations	a) Design (to the extent specified in the Contract), execute and complete the Works in accordance with the Contract and with the Engineer's instructions and remedy any defects in the Works. b) Provide the required Plant and Contractor's Documents specified in the Contract and all Contractor's Personnel, Goods, consumables and other things and services, whether of a temporary or permanent nature. c) Be responsible for the adequacy, stability and safety of all Site operations and of all methods of construction. d) Be responsible for all Contractor's Documents, Temporary Works, and such design of each item of Plant and Materials as is required for the item to be in accordance with the Contract. e) Submit details of the arrangements and methods proposed for the execution of the Works. f) Design any part of the Permanent Works if specified in the Contract. g) Submit to the Engineer the "as-built" documents and operation and maintenance manuals prior to the Tests on Completion.	None.	None.

CLAUSE	OBLIGATIONS	TIME FRAME	SPECIFIC CONSEQUENCES OF NON-COMPLIANCE
4.2 Performance Security	a) Obtain a Performance Security for proper performance and deliver to the Employer. b) Ensure that the Performance Security is valid and enforceable until the Contractor has executed and completed the Works and remedied any defects. c) Extend the validity of the Performance Security until the Works have been completed and any defects have been remedied.	a) Within 28 days after receiving the Letter of Acceptance. b) None. c) As required.	a) None. b) None. c) Employer may claim the full amount of the Performance Security.
4.3 Contractor's Representative	a) Appoint the Contractor's Representative and give him all authority necessary to act on the Contractor's behalf under the Contract. b) Submit to the Engineer for consent, the name and particulars of the person the Contractor proposes to appoint. c) If consent is withheld or subsequently revoked, or if the appointed person fails to act, submit the name and particulars of another suitable person for such appointment. d) Not, without the prior consent of the Engineer, revoke the appointment of the Contractor's Representative or appoint a replacement. e) If the Contractor's Representative is to be temporarily absent from the Site, appoint a suitable replacement and notify the Engineer.	a) None. b) Prior to the Commencement Date. c) None. d) None. e) None.	None.
4.4 Subcontractors	a) Not to subcontract the whole of the Works. b) Be responsible for the acts or defaults of any Subcontractor, his agents or employees. c) Obtain prior consent of the Engineer to proposed Subcontractors, not named in the Contract. d) Give the Engineer notice of the intended date of the commencement of each Subcontractor's work, and of the commencement of such work on the Site.	a) None. b) None. c) None. d) Not less than 28 days.	None.

THE OBLIGATIONS OF THE CONTRACTOR

Red Book

THE OBLIGATIONS OF THE CONTRACTOR (continued)

CLAUSE	OBLIGATIONS	TIME FRAME	SPECIFIC CONSEQUENCES OF NON-COMPLIANCE
4.5 Assignment of Benefit of Subcontract	If instructed, assign the benefit of subcontract obligations which extend beyond the Defects Notification Period to the Employer.	None.	None.
4.6 Co-operation	a) Allow appropriate opportunities for carrying out work to the Employer's Personnel, any other contractors employed by the Employer and the personnel of any legally constituted public authorities. b) Submit such documents which require the Employer to give to the Contractor possession of any foundation, structure, plant or means of access.	a) None. b) In the time and manner stated in the Specification.	None.
4.7 Setting Out	a) Set out the Works in relation to original points, lines and levels of reference specified in the Contract or notified by the Engineer. b) Give notice to the Engineer if the Contractor suffers delay and/or incurs Cost as a result of error in the items of reference.	a) None. b) As soon as practicable and not later than 28 days after the Contractor became aware, or should have become aware of the event or circumstance.	a) None. b) Loss of entitlement to an extension to the Time for Completion and additional payment (Sub-Clause 20.1).
4.8 Safety Procedures	a) Comply with all applicable safety regulations. b) Take care for the safety of all persons entitled to be on the Site. c) Use reasonable efforts to keep the Site and Works clear of unnecessary obstruction. d) Provide fencing, lighting, guarding and watching of the Works. e) Provide any Temporary Works, which may be necessary for the use and protection of the public and of owners and occupiers of adjacent land.	None.	None.

CLAUSE	OBLIGATIONS	TIME FRAME	SPECIFIC CONSEQUENCES OF NON-COMPLIANCE
4.9 Quality Assurance	Institute a quality assurance system and submit details to the Engineer.	Before each design and execution stage is commenced.	None.
4.12 Unforeseeable Physical Conditions	a) Give notice of adverse physical conditions. b) Continue executing the Works, using such proper and reasonable measures as are appropriate for the physical conditions. c) Comply with any instructions which the Engineer may give. d) Give a further notice if the Contractor suffers delay and/or incurs Cost due to unforeseen physical conditions.	a) As soon as practicable. b) None. c) None. d) As soon as practicable and not later than 28 days after the Contractor became aware, or should have become aware of the event or circumstance (Sub-Clause 20.1).	a) None. b) None. c) None. d) Loss of entitlement to an extension to the Time for Completion and additional payment (Sub-Clause 20.1).
4.13 Rights of Way and Facilities	a) Bear all costs and charges for special and/or temporary rights-of-way. b) Obtain any additional facilities outside the Site which the Contractor may require for the purposes of the Works.	None.	None.
4.14 Avoidance of Interference	Not to interfere with the convenience of the public, or the access to and use and occupation of all roads and footpaths.	None.	None.
4.15 Access Route	a) Use reasonable efforts to prevent any road or bridge from being damaged. b) Be responsible for any maintenance which may be required for the use of access routes. c) Provide all necessary signs or directions along access routes. d) Obtain any permission which may be required from the relevant authorities for use of routes, signs and directions.	None.	None.

THE OBLIGATIONS OF THE CONTRACTOR

THE OBLIGATIONS OF THE CONTRACTOR (continued)

CLAUSE	OBLIGATIONS	TIME FRAME	SPECIFIC CONSEQUENCES OF NON-COMPLIANCE
4.16 Transport of Goods	a) Give the Engineer notice of the date on which any Plant or a major item of other Goods will be delivered to the Site. b) Be responsible for packing, loading, transporting, receiving, unloading, storing and protecting all Goods and other things required for the Works.	a) 21 days before delivery. b) None.	None.
4.17 Contractor's Equipment	a) Be responsible for all Contractor's Equipment. b) Not to remove from the Site any major items of Contractor's Equipment without the consent of the Engineer.	None.	None.
4.18 Protection of the Environment	a) Take all reasonable steps to protect the environment and to limit damage and nuisance to people and property. b) Ensure that emissions, surface discharges and effluent shall not exceed the values indicated in the Specification or prescribed by applicable Laws.	None.	None.
4.19 Electricity, Water and Gas	a) Be responsible for the provision of all power, water and other services. b) Provide any apparatus necessary for use of services as may be available on the Site and for measuring the quantities consumed. c) Pay the Employer for the use of services available on the Site.	None.	None.

CLAUSE	OBLIGATIONS	TIME FRAME	SPECIFIC CONSEQUENCES OF NON-COMPLIANCE
4.20 Employer's Equipment and Free-Issue Materials	a) Be responsible for the Employers' Equipment when used by the Contractor. b) Pay the Employer for the use of the Employers' Equipment. c) Inspect free-issue materials. d) Give notice of any shortage, defect or default in the free-issue materials.	a) None. b) None. c) None. d) Promptly.	None.
4.21 Progress Reports	Prepare and submit monthly progress reports.	Monthly, within 7 days of the period to which the report relates.	None.
4.22 Security of the Site	Keep unauthorised persons off the Site.	None.	None.
4.23 Contractor's Operations on Site	a) Confine operations to the Site, and to any additional areas agreed as working areas. b) Take all necessary precautions to keep Contractor's Equipment and Contractor's Personnel within the Site and any agreed working areas. c) Keep the Site free from all unnecessary obstruction. d) Store or dispose of any Contractor's Equipment or surplus materials. e) Clear away and remove from the Site any wreckage, rubbish and Temporary Works. f) Leave the Site and the Works in a clean and safe condition.	None.	None.
4.24 Fossils	a) Take reasonable precautions to prevent Contractor's Personnel or other persons from removing or damaging fossils, coins, articles of value or antiquity, structures and other remains or items of geological or archaeological interest. b) Give notice of the finding of such items. c) Give further notice if the Contractor suffers delay and/or incurs Cost as a result of such items.	a) None. b) Upon discovery. c) As soon as practicable and not later than 28 days after the Contractor became aware, or should have become aware of the event or circumstance (Sub-Clause 20.1).	a) None. b) None. c) Loss of entitlement to an extension to the Time for Completion and additional payment (Sub-Clause 20.1).

THE OBLIGATIONS OF THE CONTRACTOR

CLAUSE	OBLIGATIONS	TIME FRAME	SPECIFIC CONSEQUENCES OF NON-COMPLIANCE
5 Nominated Subcontractors			
5.3 Payments to Nominated Subcontractors	Pay to the nominated Subcontractor the amounts which the Engineer certifies.	None.	None.
6 Staff and Labour			
6.1 Engagement of Staff and Labour	Make arrangements for the engagement of all staff and labour, local or otherwise, and for their payment, housing, feeding and transport.	None.	None.
6.2 Rates of Wages and Conditions of Labour	Pay rates of wages, and observe conditions of labour which are not lower than those established for the trade or industry where the work is carried out.	None.	None.
6.3 Persons in the Service of Employer	Not recruit, or attempt to recruit, staff and labour from amongst the Employer's Personnel.	None.	None.
6.4 Labour Laws	a) Comply with all the relevant labour Laws. b) Require employees to obey all applicable Laws.	None.	None.
6.5 Working Hours	Obtain the consent of the Engineer if working outside the normal working hours.	None.	None.

CLAUSE	OBLIGATIONS	TIME FRAME	SPECIFIC CONSEQUENCES OF NON-COMPLIANCE
6.6 Facilities for Staff and Labour	a) Provide and maintain all necessary accommodation and welfare facilities for the Contractor's Personnel. b) Provide facilities for the Employer's Personnel as stated in the Specification. c) Not permit any of the Contractor's Personnel to maintain any temporary or permanent living quarters within the structures forming part of the Permanent Works.	None.	None.
6.7 Health and Safety	a) Take all reasonable precautions to maintain the health and safety of the Contractor's Personnel. b) Ensure that medical staff, first aid facilities, sick bay and ambulance service are available at all times and that suitable arrangements are made for all necessary welfare and hygiene requirements and for the prevention of epidemics. c) Appoint an accident prevention officer and whatever is required by this person to exercise this responsibility and authority. d) Send to the Engineer details of any accident. e) Maintain records and make reports concerning health, safety, welfare and damage to property.	a) None. b) None. c) None. d) As soon as practicable after its occurrence. e) None.	None.
6.8 Contractor's Superintendence	Provide all necessary superintendence to plan, arrange, direct, manage, inspect and test the work.	None.	None.
6.10 Records of Contractor's Personnel and Equipment	Submit to the Engineer, details showing the number of each class of Contractor's Personnel and of each type of Contractor's Equipment on the Site.	Each calendar month.	None.
6.11 Disorderly Conduct	Take all reasonable precautions to prevent any unlawful, riotous or disorderly conduct by, or amongst the Contractor's Personnel.	None.	None.

THE OBLIGATIONS OF THE CONTRACTOR (continued)

CLAUSE	OBLIGATIONS	TIME FRAME	SPECIFIC CONSEQUENCES OF NON-COMPLIANCE
7 Plant, Materials and Workmanship			
7.1 Manner of Execution	Carry out the manufacture of Plant, the production and manufacture of Materials and all other execution of the Works.	None.	None.
7.2 Samples	Submit samples of Materials and relevant information, to the Engineer for consent.	Prior to using the Materials.	None.
7.3 Inspection	a) Give the Employer's Personnel full opportunity to carry out inspections. b) Give notice to the Engineer to inspect.	a) None. b) Whenever any work is ready and before it is covered up, put out of sight or packaged for storage or transport.	a) None. b) Contractor obliged to uncover the work, reinstate and make good at the Contractor's cost.
7.4 Testing	a) Provide everything necessary to carry out the specified tests. b) Agree with the Engineer the time and place for the testing. c) Give notice if the Contractor suffers delay and/or incurs Cost as a result of complying with instructions or a delay for which the Employer is responsible. d) Forward to the Engineer certified reports of the tests.	a) None. b) None. c) As soon as practicable and not later than 28 days after the Contractor became aware, or should have become aware of the event or circumstance (Sub-Clause 20.1). d) Promptly.	a) None. b) None. c) Loss of entitlement to an extension to the Time for Completion and additional payment (Sub-Clause 20.1). d) None.
7.5 Rejection	Make good defects notified by the Engineer.	Promptly.	None.
7.6 Remedial Work	Comply with the instructions of the Engineer with regard to remedial work.	Within a reasonable time as specified in the instruction or immediately if urgency is specified.	Contractor shall pay costs incurred by the Employer in engaging other persons to carry out the work.

CLAUSE	OBLIGATIONS	TIME FRAME	SPECIFIC CONSEQUENCES OF NON-COMPLIANCE
7.8 Royalties	Pay all royalties, rents and other payments for natural Materials obtained from outside the Site and disposal of surplus materials.	None.	None.
8 Commencement, Delays and Suspension			
8.1 Commencement of Works	Commence the execution of the Works and proceed with the Works with due expedition and without delay.	As soon as is reasonably practicable after the Commencement Date.	None.
8.2 Time for Completion	Complete the whole of the Works and each Section within the times specified in the Contract.	None.	Contractor shall pay delay damages to the Employer (Sub-Clause 8.7).
8.3 Programme	a) Submit a detailed time programme. b) Submit a revised programme. c) Proceed in accordance with the programme. d) Give notice to the Engineer of specific probable future events or circumstances which may adversely affect the work, increase the Contract Price or delay the execution of the Works. e) Submit a revised programme on receiving a notice from the Engineer that a programme fails to comply with the Contract or to be consistent with actual progress.	a) Within 28 days after receiving the notice of commencement. b) Whenever the previous programme is inconsistent with actual progress or with the Contractor's obligations. c) None. d) Promptly. e) None.	None.
8.4 Extension of Time for Completion	Give notice to the Engineer if the Contractor considers himself to be entitled to an extension of the Time for Completion.	As soon as practicable and not later than 28 days after the Contractor became aware, or should have become aware of the event or circumstance (Sub-Clause 20.1).	Loss of entitlement to an extension to the Time for Completion (Sub-Clause 20.1).
8.6 Rate of Progress	Adopt revised methods in order to expedite progress and complete within the Time for Completion.	None.	None.

THE OBLIGATIONS OF THE CONTRACTOR (continued)

CLAUSE	OBLIGATIONS	TIME FRAME	SPECIFIC CONSEQUENCES OF NON-COMPLIANCE
8.7 Delay Damages	Pay delay damages in the case of failure to comply with the Time for Completion.	None.	None.
8.8 Suspension of Work	Protect, store and secure such part or the Works in the case of an instruction to suspend the Works.	None.	None.
8.9 Consequences of Suspension	Give notice to the Engineer if the Contractor suffers delay and/or incurs cost as a result of complying with the Engineer's instructions under Sub-Clause 8.8.	As soon as practicable and not later than 28 days after the Contractor became aware, or should have become aware, of the event or circumstance (Sub-Clause 20.1).	Loss of entitlement to an extension to the Time for Completion and additional payment (Sub-Clause 20.1).
8.12 Resumption of Work	a) Jointly examine the Works and the Plant and Materials affected by the suspension with the Engineer. b) Make good any deterioration, defect or loss.	None.	None.
9 Tests on Completion			
9.1 Contractor's Obligations	a) Carry out the Tests on Completion. b) Give to the Engineer notice of the date after which the Contractor will be ready to carry out each of the Tests on Completion. c) Submit a certified report of the results of the Tests to the Engineer.	a) After providing the documents in accordance with Sub-Clause 4.1(d). b) Not less than 21 days. c) As soon as the Works or a Section have passed the Tests on Completion.	None.
9.2 Delayed Tests	Carry out the Tests if the Engineer gives notice of undue delay.	Within 21 days of the Engineer's notice.	The Employer's Personnel may proceed with the tests at the Contractor's cost.

CLAUSE	OBLIGATIONS	TIME FRAME	SPECIFIC CONSEQUENCES OF NON-COMPLIANCE
10 Employer's Taking Over			
10.2 Taking Over of Parts of the Works	a) Carry out any outstanding Tests on Completion. b) Give notice of costs incurred as a result of the Employer taking over and/or using a part of the Works.	a) As soon as practicable. b) As soon as practicable and not later than 28 days after the Contractor became aware, or should have become aware of the event or circumstance (Sub-Clause 20.1).	a) None. b) Loss of entitlement to additional payment (Sub-Clause 20.1).
10.3 Interference with Tests on Completion	a) In the case of prevention from carrying out the tests, carry out any outstanding Tests on Completion. b) Give notice if the Contractor suffers delay and/or incurs Cost as a result of interference with Tests on Completion.	a) As soon as practicable. b) As soon as practicable and not later than 28 days after the Contractor became aware, or should have become aware of the event or circumstance (Sub-Clause 20.1).	c) None. d) Loss of entitlement to an extension to the Time for Completion and additional payment (Sub-Clause 20.1).
11 Defects Liability			
11.1 Completion of Outstanding Work and Remedying Defects	a) Complete any work which is outstanding on the date stated in a Taking-Over Certificate. b) Execute all work required to remedy defects or damage.	a) Within such reasonable time as is instructed by the Engineer. b) On or before the expiry date of the Defects Notification Period.	a) The Employer may carry out the work himself, at the Contractor's cost (Sub-Clause 11.4). b) A reduction in the Contract Price may be made (Sub-Clause 11.4).
11.8 Contractor to Search	If required by the Engineer, search for the cause of any defect.	None.	None.

THE OBLIGATIONS OF THE CONTRACTOR (continued)

CLAUSE	OBLIGATIONS	TIME FRAME	SPECIFIC CONSEQUENCES OF NON-COMPLIANCE
11.11 Clearance of Site	Remove any remaining Contractor's Equipment, surplus material, wreckage, rubbish and Temporary Works from the Site.	Within 28 days of receipt of the Performance Certificate.	a) The Employer may sell or otherwise dispose of any remaining items. b) Employer entitled to recover costs of disposal.
12 Measurement and Evaluation			
12.1 Works to be Measured	a) Assist the Engineer in making the measurement. b) Supply any particulars requested by the Engineer. c) Examine and agree the records with the Engineer and sign the same when agreed. d) In the case of disagreement with the records, give notice to the Engineer.	a) Promptly. b) None. c) As and when requested. c) Within 14 days after examination.	The Engineer's records shall be accepted as accurate.
12.4 Omissions	In the case where the Contractor will incur cost or not be adequately compensated for omitted work, give notice to the Engineer with supporting particulars.	None.	None.
13 Variations and Adjustments			
13.1 Right to Vary	a) Execute and be bound by each Variation. b) Not make any alteration and/or modification of the Permanent Works, unless and until the Engineer instructs or approves a Variation.	None.	None.

CLAUSE	OBLIGATIONS	TIME FRAME	SPECIFIC CONSEQUENCES OF NON-COMPLIANCE
13.3 Variation Procedure	a) Respond in writing to a request for a proposal. b) Not delay any work whilst awaiting a response. c) Acknowledge receipt of Variation instructions.	a) As soon as practicable. b) None. c) None.	None.
13.5 Provisional Sums	Produce quotations, invoices, vouchers and accounts or receipts in substantiation of the amounts paid to nominated Subcontractors.	When required by the Engineer.	None.
13.6 Daywork	a) Submit quotations to the Engineer. b) Submit invoices, vouchers and accounts or receipts for any Goods. c) Deliver to the Engineer statements which include the details of the resources used in executing the previous day's work. d) Submit priced statements of these resources.	a) Before ordering Goods for the work to be executed on a Daywork basis. b) When applying for payment. c) Each day. d) Prior to their inclusion in the next Statement under Sub-Clause 14.3.	None.
13.7 Adjustments for Changes in Legislation	Give notice if the Contractor suffers delay and/or incurs Cost as a result of changes in legislation.	As soon as practicable and not later than 28 days after the Contractor became aware, or should have become aware of the event or circumstance (Sub-Clause 20.1).	Loss of entitlement to an extension to the Time for Completion and additional payment (Sub-Clause 20.1).
14 Contract Price and Payment			
14.1 The Contract Price	a) Pay all taxes, duties and fees to be paid under the Contract. b) Submit to the Engineer a proposed breakdown of each lump sum price in the Schedules.	a) None. b) Within 28 days after the Commencement Date.	None.
14.2 Advance Payment	a) Submit an advance payment guarantee. b) Extend the validity of the guarantee until the advance payment has been repaid.	None.	a) Employer is not obliged to make the advance payment. b) None.

THE OBLIGATIONS OF THE CONTRACTOR (continued)

CLAUSE	OBLIGATIONS	TIME FRAME	SPECIFIC CONSEQUENCES OF NON-COMPLIANCE
14.3 Application for Interim Payment Certificates	Submit a Statement in six copies, showing in detail the amounts to which the Contractor considers himself to be entitled.	After the end of each month.	No obligation on the Engineer to certify payment (Sub-Clause 14.6).
14.4 Schedule of Payments	In the case that the Contract does not include a schedule of payments, submit non-binding estimates of the payments expected to become due.	a) First estimate within 42 days after the Commencement Date. b) Revised estimates at quarterly intervals.	None.
14.10 Statement at Completion	Submit a Statement at completion.	Within 84 days after receiving the Taking-Over Certificate for the Works.	None.
14.11 Application for Final Payment Certificate	a) Submit a draft Final Statement. b) Submit such further information as the Engineer may reasonably require. c) Prepare and submit the final statement as agreed with the Engineer.	a) Within 56 days after receiving the Performance Certificate. b) None. c) None.	a) None. b) None. c) The Engineer will certify an amount that he fairly determines to be due.
14.12 Discharge	Submit a written discharge.	When submitting the Final Statement.	None.

CLAUSE	OBLIGATIONS	TIME FRAME	SPECIFIC CONSEQUENCES OF NON-COMPLIANCE
15 Termination by Employer			
15.2 Termination by Employer	a) In the case of a notice of termination being served, leave the Site and deliver any required Goods, Contractor's Documents and other design documents to the Engineer. b) Use best efforts to comply with any reasonable instructions included in the notice. c) Arrange for the removal of Equipment and Temporary Works.	a) None. b) Immediately. c) Promptly.	a) None. b) None. c) Items may be sold by the Employer.
15.5 Employer's Entitlement to Termination	In the case of a notice of termination, cease all further work, hand over Contractor's Documents, Plant, Materials and other work and remove all other Goods from the Site (Sub-Clause 16.3).	28 days from the Employer's notice or return of the Performance Security, whichever is the later.	None.
16 Suspension and Termination by Contractor			
16.1 Contractor's Entitlement to Suspend Work	a) Give notice if the Contractor intends to suspend work or reduce the rate of work. b) Resume normal working when the Employer's obligations have been met. c) Give further notice if the Contractor suffers delay and/or incurs Cost as a result of suspending work or reducing the rate of work.	a) 21 days before the intended suspension or reduction in the rate of work. b) As soon as is reasonably practicable. c) As soon as practicable and not later than 28 days after the Contractor became aware, or should have become aware of the event or circumstance (Sub-Clause 20.1).	a) None. b) None. c) Loss of entitlement to an extension to the Time for Completion and additional payment (Sub-Clause 20.1).
16.2 Termination by Contractor	Give notice of intention to terminate.	14 days before the intended termination date.	None.
16.3 Cessation of Work and Removal of Contractor's Equipment	In the case of a notice of termination, cease all further work, hand over Contractor's Documents, Plant, Materials and other work and remove all other Goods from the Site.	After the notice has taken effect.	None.

THE OBLIGATIONS OF THE CONTRACTOR

Red Book

THE OBLIGATIONS OF THE CONTRACTOR (continued)

CLAUSE	OBLIGATIONS	TIME FRAME	SPECIFIC CONSEQUENCES OF NON-COMPLIANCE
17 Risk and Responsibility			
17.1 Indemnities	Indemnify and hold harmless the Employer's Personnel and their respective agents against and from all claims, damages, losses and expenses in respect of bodily injury, sickness, disease, death, damage to or of loss of property by reason of the Contractor's design, the execution and completion of the Works.	None.	None.
17.2 Contractor's Care	a) Take full responsibility for the care of the Works and Goods. b) Take responsibility for the care of any work which is outstanding on the date stated in a Taking-Over Certificate. c) Rectify loss or damage if any loss or damage happens to the Works, Goods or Contractor's Documents.	a) From the Commencement Date until the Taking-Over Certificate is issued. b) Until the outstanding work has been completed. c) None.	None.
17.4 Consequences of Employer's Risks	a) Give notice in the case of an Employer's risk event which results in loss or damage. b) Rectify the loss or damage as required by the Engineer. c) Give further notice if the Contractor suffers delay and/or incurs Cost as a result of rectifying loss or damage caused by Employers Risks.	a) Promptly. b) None. c) As soon as practicable and not later than 28 days after the Contractor became aware, or should have become aware of the event or circumstance (Sub-Clause 20.1).	a) None. b) None. c) Loss of entitlement to an extension to the Time for Completion and additional payment (Sub-Clause 20.1).

CLAUSE	OBLIGATIONS	TIME FRAME	SPECIFIC CONSEQUENCES OF NON-COMPLIANCE
17.5 Intellectual and Industrial Property Rights	a) Give notice of any claim under this clause. b) Indemnify and hold the Employer harmless against and from any other claim which arises out of, or in relation to the manufacture, use, sale or import of any Goods, or any design for which the Contractor is responsible. c) If requested by the Employer, assist in contesting the claim. d) Not make any admission which might be prejudicial to the Employer.	a) Within 28 days of receiving a claim. b) None. c) None. d) None.	a) Waiver of right to indemnity. b) None. c) None. d) None.
18 Insurance			
18.1 General Requirements for Insurances	a) Wherever the Contractor is the insuring Party, effect and maintain the insurances in terms consistent with any terms agreed by the Parties before the date of the Letter of Acceptance. b) Act under the policy on behalf of any additional joint insured parties. c) Submit evidence to the Employer that the insurances have been effected and copies of the policies. d) Submit evidence of payment of premiums. e) Inform the insurers of any relevant changes to the execution of the Works and ensure that insurance is maintained. f) Not make any material alteration to the terms of any insurance without approval of the Employer.	a) Within the periods stated in the Contract Data. b) None. c) Within the periods stated in the Appendix to Tender. d) Upon payment of premium. e) As appropriate. f) None.	Employer may effect the insurances and recover the cost from the Contractor.
18.4 Insurance for Contractor's Personnel	Effect and maintain insurance against injury, sickness, disease or death of any person employed by the Contractor, or any other of the Contractor's Personnel.	From the time that personnel are assisting in the execution of the Works.	Employer may effect the insurances and recover the cost from the Contractor (Sub-Clause 18.1).

THE OBLIGATIONS OF THE CONTRACTOR (continued)

CLAUSE	OBLIGATIONS	TIME FRAME	SPECIFIC CONSEQUENCES OF NON-COMPLIANCE
19 Force Majeure			
19.2 Notice of Force Majeure	Give notice in the case that the Contractor is, or will be prevented from performing any of its obligations under the Contract by Force Majeure.	Within 14 days after the Contractor became aware, or should have become aware of the relevant event or circumstance constituting Force Majeure.	Contractor shall not be excused performance of the obligations.
19.3 Duty to Minimise Delay	a) Use all reasonable endeavours to minimise any delay in the performance of the Contract as a result of Force Majeure. b) Give notice when the effects of the Force Majeure cease.	None.	None.
20 Claims, Disputes and Arbitration			
20.1 Contractor's Claims	a) Give notice if the Contractor considers himself to be entitled to any extension of the Time for Completion and/or any additional payment. b) Submit any other notices which are required by the Contract and supporting particulars of the claim. c) Keep such contemporary records as may be necessary to substantiate any claim and permit the Engineer to inspect all the records. d) Send to the Engineer, a fully detailed claim. e) Send further interim claims if the event or circumstance giving rise to the claim has a continuing effect. f) Send a final claim.	a) As soon as practicable and not later than 28 days after the Contractor became aware, or should have become aware of the event or circumstance. b) None. c) None. d) Within 42 days after the Contractor became aware (or should have become aware) of the event or circumstance giving rise to the claim. e) At monthly intervals. f) Within 28 days after the end of the effects resulting from the event or circumstance.	a,b,d) Loss of entitlement to an extension to the Time for Completion and additional payment. c,e,f) The Employer will take account of the extent to which the failure has prevented or prejudiced proper investigation of the claim.

CLAUSE	OBLIGATIONS	TIME FRAME	SPECIFIC CONSEQUENCES OF NON-COMPLIANCE
20.2 Appointment of the Dispute Adjudication Board	a) Jointly appoint the DAB (Dispute Adjudication Board). b) Mutually agree the terms of remuneration for the DAB. c) Not consult the DAB on any matter without the agreement of the Employer. d) Not to act alone in the termination of any member of the DAB.	a) By the date stated in the Appendix to Tender. b) None. c) None. d) None.	a) The appointing entity or official named in the Appendix to Tender shall appoint (Sub-Clause 20.3). b) None. c) None. d) None.
20.4 Obtaining Dispute Adjudication Board's Decision	a) Make available to the DAB, additional information, access to the Site and appropriate facilities as the DAB may require. b) Give effect to a DAB decision unless and until it is revised, in an amicable settlement or an arbitral award. c) Continue to proceed with the Works in accordance with the Contract.	a) Promptly. b) Promptly. c) None.	a) None. b&c) The matter may be referred to arbitration (Sub-Clause 20.7).
20.5 Amicable Settlement	In the case of a notice of dissatisfaction being issued, attempt to settle the dispute amicably.	Within 56 days of the notice.	Arbitration may be commenced.

THE OBLIGATIONS OF THE CONTRACTOR (continued)

CLAUSE	OBLIGATIONS	TIME FRAME	SPECIFIC CONSEQUENCES OF NON-COMPLIANCE
GENERAL CONDITIONS OF DISPUTE ADJUDICATION AGREEMENT			
2 General Provisions	Give notice to the DAB Member that the Dispute Adjudication Agreement has taken effect.	On the Commencement Date, or upon all parties signing the Dispute Adjudication Agreement whichever is the later.	None.
5 General Obligations of the Employer and the Contractor	a) Not to request advice from, or consult with the Member regarding the Contract, otherwise than in the normal course of the DAB's activities. b) In the case of the DAB Member being required to make a site visit or attend a hearing, provide appropriate security for a sum equivalent to the reasonable expenses to be incurred by the Member (this may be undertaken by the Employer).	None.	None.
6 Payment	a) Pay the DAB fees. b) Apply to the Employer for reimbursement of one-half of the DAB invoices by way of the Statements.	a) Within 56 calendar days after receiving each invoice. b) None	a) Employer may pay the fees and be entitled to reimbursement of fees, plus financing charges. The DAB Member may suspend services or resign the appointment. b) None

CLAUSE	OBLIGATIONS	TIME FRAME	SPECIFIC CONSEQUENCES OF NON-COMPLIANCE
Annex – Procedural Rules			
2.	Jointly agree the timing of and agenda for each site visit by the DAB.	None.	Timing and agenda shall be decided by the DAB.
3.	a) Attend site visits by the DAB. b) Co-operate with the Employer in co-ordinating site visits by the DAB.	None.	None.
4.	a) Furnish to each DAB member, one copy of all documents which the DAB may request. b) Copy the Employer on all communications between the DAB and the Contractor.	None.	None.

THE OBLIGATIONS OF THE ENGINEER

CLAUSE	OBLIGATIONS	TIME FRAME	SPECIFIC CONSEQUENCES OF NON-COMPLIANCE
GENERAL CONDITIONS			
1 General Provisions			
1.3 Communications	Not to unreasonably withhold approvals, certificates, consents and determinations.	None.	None.
1.5 Priority of Documents	In the case that an ambiguity or discrepancy is found in the Contract documents, issue any necessary clarification or instruction.	None.	None.
1.9 Delayed Drawings or Instructions	In the case of a notice and claim for delay or cost being received, respond to the claim and agree or determine the matters.	Respond within 42 days after receiving a claim or any further particulars supporting a previous claim (Sub-Clause 20.1).	None.
2 The Employer			
2.1 Right of Access to the Site	In the case of a notice and claim for delay or cost being received, respond to the claim and agree or determine the matters.	Respond within 42 days after receiving a claim or any further particulars supporting a previous claim (Sub-Clause 20.1).	None.
2.5 Employer's Claims	a) In the case that the Employer considers himself to be entitled to any payment, give notice and particulars to the Contractor (the Employer may also undertake this action). b) Agree or determine the matters.	a) As soon as practicable after the Employer became aware of the event or circumstances giving rise to the claim. b) None.	None.

CLAUSE	OBLIGATIONS	TIME FRAME	SPECIFIC CONSEQUENCES OF NON-COMPLIANCE
3 The Engineer			
3.1 Engineer's Duties and Authority	a) Carry out the duties assigned in the Contract. b) Provide staff that are suitably qualified engineers and other professionals who are competent to carry out these duties. c) Obtain the approval of the Employer before exercising any authority specified in the Particular Conditions.	None.	a) None. b) None. c) The Employer shall be deemed to have given approval.
3.2 Delegation by the Engineer	a) In the case of delegation of the Engineer's authority, delegate such authority in writing. b) Not to delegate the authority to determine any matter in accordance with Sub-Clause 3.5 *[Determinations]*. c) In the case that the Contractor questions any determination or instruction of an assistant and refers the matter, confirm, reverse or vary the determination or instruction.	a) None. b) None. c) Promptly.	None.
3.3 Instructions of the Engineer	Wherever practical give instructions in writing.	None.	None.
3.5 Determinations	a) Consult with each Party in an endeavour to reach an agreement. b) If agreement is not achieved, make a fair determination in accordance with the Contract, taking due regard of all relevant circumstances. c) Give notice to both Parties of each agreement or determination, with supporting particulars.	None.	None.

THE OBLIGATIONS OF THE ENGINEER

Red Book

THE OBLIGATIONS OF THE ENGINEER (continued)

CLAUSE	OBLIGATIONS	TIME FRAME	SPECIFIC CONSEQUENCES OF NON-COMPLIANCE
4 The Contractor			
4.3 Contractor's Representative	Respond to the Contractor's request for consent to the appointment of the Contractor's Representative.	None.	None.
4.4 Subcontractors	Respond to the Contractor's requests for consent for proposed Subcontractors.	None.	None.
4.7 Setting Out	In the case of a notice and claim for delay or cost being received, respond to the claim and agree or determine the matters.	Respond within 42 days after receiving a claim or any further particulars supporting a previous claim (Sub-Clause 20.1).	None.
4.12 Unforeseeable Physical Conditions	a) In the case of a notice of unforeseen physical conditions, inspect the physical conditions. b) In the case of a notice and claim for delay or Cost being received, respond to the claim and agree or determine the matters.	a) None. b) Respond within 42 days after receiving a claim or any further particulars supporting a previous claim (Sub-Clause 20.1).	None.
4.19 Electricity, Water and Gas	Agree or determine the quantities and amounts due to the Employer for the Contractor's consumption of electricity, water and gas.	None.	None.
4.20 Employer's Equipment and Free-Issue Material	Agree or determine the amounts due to the Employer for the Contractor's use of Employer's Equipment.	None.	None.

CLAUSE	OBLIGATIONS	TIME FRAME	SPECIFIC CONSEQUENCES OF NON-COMPLIANCE
4.23 Contractor's Operations on Site	Cooperate with the Contractor to agree additional working areas outside the Site.	None.	None.
4.24 Fossils	a) Give instructions for dealing with fossils, coins, articles of value or antiquity, and structures and other remains or items of geological or archaeological interest found on the Site. b) In the case of a notice and claim for delay or Cost being received, respond to the claim and agree or determine the matters.	a) None. b) Respond within 42 days after receiving a claim or any further particulars supporting a previous claim (Sub-Clause 20.1).	None.
5 Nominated Subcontractors			
5.3 Payments to Nominated Subcontractors	Certify the amounts due to nominated Subcontractors.	None.	None.
6 Staff and Labour			
6.10 Records of Contractor's Personnel and Equipment	Cooperate with the Contractor to approve a form to record the number of each class of Contractor's Personnel and each type of Contractor's Equipment on the Site.	None.	None.
7 Plant, Materials and Workmanship			
7.3 Inspection	Examine, inspect, measure and test the materials and workmanship.	Without unreasonable delay (or promptly give notice that inspection is not required).	None.

THE OBLIGATIONS OF THE ENGINEER

THE OBLIGATIONS OF THE ENGINEER (continued)

CLAUSE	OBLIGATIONS	TIME FRAME	SPECIFIC CONSEQUENCES OF NON-COMPLIANCE
7.4 Testing	a) Agree with the Contractor the time and place for specified testing. b) Give the Contractor notice of intention to attend the tests. c) In the case of a notice and claim for delay or Cost being received, respond to the claim and agree or determine the matters. d) Endorse the Contractor's test certificate, or issue a certificate confirming that the tests have been passed.	a) None. b) Not less than 24 hours. c) Respond within 42 days after receiving a claim or any further particulars supporting a previous claim (Sub-Clause 20.1). d) Promptly.	a) None. b) If the Engineer does not attend, the Contractor may proceed and the tests shall be deemed to have been made in Engineer's presence. c) None. d) None.
7.5 Rejection	In the case of Plant, Materials or workmanship being found to be defective or not in accordance with the Contract, give notice of rejection.	None.	None.
8 Commencement, Delays and Suspension			
8.1 Commencement of Works	Give the Contractor notice of the Commencement Date.	Not less than 7 days before the Commencement Date and within 42 days after the Contractor receives the Letter of Acceptance.	None.
8.3 Programme	In the case that a programme does not comply with the Contract, give notice to the Contractor.	Within 21 days after receiving the programme.	Contractor shall proceed in accordance with the programme.
8.4 Extension of Time for Completion	In the case of a notice and claim for delay being received, respond to the claim and agree or determine the matters.	Respond within 42 days after receiving a claim or any further particulars supporting a previous claim (Sub-Clause 20.1).	None.

CLAUSE	OBLIGATIONS	TIME FRAME	SPECIFIC CONSEQUENCES OF NON-COMPLIANCE
8.8 Suspension of Work	In the case of a notice and claim for delay or Cost being received, respond to the claim and agree or determine the matters.	Respond within 42 days after receiving a claim or any further particulars supporting a previous claim (Sub-Clause 20.1).	None.
8.11 Prolonged Suspension	In the case that the Contractor requests permission to proceed after 84 days of suspension, respond to the Contractor's request.	Within 28 days of the request.	Contractor may treat the suspension as an omission or give notice of termination.
8.12 Resumption of Work	Jointly examine the Works and the Plant and Materials affected by the suspension.	After permission or instruction to proceed is given.	None.
9 Tests on Completion			
9.1 Contractor's Obligations	Make allowances for the effect of any use of the Works by the Employer on the performance or other characteristics of the Works.	None.	None.
10 Employer's Taking Over			
10.1 Taking Over of the Works and Sections	Issue the Taking-Over Certificate to the Contractor, or reject the Contractor's application, giving reasons and specifying the work required to be done.	Within 28 days after receiving the Contractor's application.	The Taking-Over Certificate shall be deemed to have been issued.
10.2 Taking Over of Parts of the Works	a) In the case that the Employer uses part of the Works and if requested by the Contractor, issue a Taking-Over Certificate for this part. b) In the case of a notice and claim for incurred Cost being received, respond to the claim and agree or determine the matters. c) Determine any reduction in delay damages as a result of a Taking-Over Certificate being issued for a part of the Works.	a) None. b) Respond within 42 days after receiving a claim or any further particulars supporting a previous claim (Sub-Clause 20.1). c) None.	None.

THE OBLIGATIONS OF THE ENGINEER (continued)

CLAUSE	OBLIGATIONS	TIME FRAME	SPECIFIC CONSEQUENCES OF NON-COMPLIANCE
10.3 Interference with Tests on Completion	a) In the case of the Contractor being prevented from carrying out Tests on Completion by the Employer, issue a Taking-Over Certificate accordingly. b) In the case of a notice and claim for Cost or delay being received, respond to the claim and agree or determine the matters.	a) None. b) Respond within 42 days after receiving a claim or any further particulars supporting a previous claim (Sub-Clause 20.1).	None.
11 Defects Liability			
11.4 Failure to Remedy Defects	In the case that the Contractor fails to remedy any defect or damage and if required by the Employer, agree or determine a reasonable reduction in the Contract Price.	None.	None.
11.8 Contractor to Search	In the case that the Contractor has searched for a defect that is found not to be the responsibility of the Contractor, agree or determine the cost of the search.	None.	None.
11.9 Performance Certificate	Issue the Performance Certificate.	Within 28 days after the latest of the expiry dates of the Defects Notification Periods or as soon thereafter as the Contractor has completed his obligations.	None.

CLAUSE	OBLIGATIONS	TIME FRAME	SPECIFIC CONSEQUENCES OF NON-COMPLIANCE
12 Measurement and Evaluation			
12.1 Works to be Measured	a) Give notice when the Engineer requires any part of the Works to be measured. b) Prepare records of measurement. c) In the case that the Contractor gives notice or disagrees with the records, review the records and either confirm or vary them.	None.	None.
12.3 Evaluation	a) Agree or determine the Contract Price by evaluating each item of work, applying the measurement and the appropriate rate or price for the item. b) In the case that new rates and prices are required and not agreed, determine a provisional rate or price for the purposes of Interim Payment Certificates.	a) None. b) Such that the item may be included in Interim Payment Certificates.	None.
12.4 Omissions	In the case of receiving a notice of cost as a result of omissions, agree or determine the cost.	None.	None.
13 Variations and Adjustments			
13.1 Right to Vary	In the case that the Contractor gives notice that the Contractor cannot readily obtain the Goods required for a Variation, cancel, confirm or vary the instruction.	None.	None.
13.2 Value Engineering	In the case that a proposal results in a reduction in the contract value, agree or determine a fee.	None.	None.

THE OBLIGATIONS OF THE ENGINEER (continued)

CLAUSE	OBLIGATIONS	TIME FRAME	SPECIFIC CONSEQUENCES OF NON-COMPLIANCE
13.3 Variation Procedure	a) Respond to the Contractor's Variation or value engineering proposals with approval, disapproval or comments. b) Issue instructions to execute Variations.	a) As soon as practicable after receiving the proposal. b) None.	a) None. b) The Contractor shall not make any alteration and/or modification of the Permanent Works (Sub-Clause 13.1).
13.5 Provisional Sums	Give instructions for the use of Provisional Sums.	None.	None.
13.6 Daywork	Sign the Contractor's Daywork Statements.	If correct, or when agreed.	None.
13.7 Adjustments for Changes in Legislation	In the case of a notice and claim for delay being received, respond to the claim and agree or determine the matters.	Respond within 42 days after receiving a claim or any further particulars supporting a previous claim (Sub-Clause 20.1).	None.
13.8 Adjustments for Changes in Cost	a) In the case that the cost indices or reference prices stated in the table of adjustment data is in doubt, make a determination. b) In the case that each current cost index is not available, determine a provisional index for the issue of Interim Payment Certificates.	a) None. b) Such that the index may be used for calculations for inclusion in the Payment Certificates.	None.
14 Contract Price and Payment			
14.2 Advance Payment	Issue an Interim Payment Certificate for the first instalment of the advance payment.	After receiving a Statement under Sub-Clause 14.3 and after the Employer receives the Performance Security and an advance payment guarantee.	None.

CLAUSE	OBLIGATIONS	TIME FRAME	SPECIFIC CONSEQUENCES OF NON-COMPLIANCE
14.3 Application for Interim Payment Certificates	Cooperate with the Contractor to agree and approve the form for the Statements.	None.	None.
14.5 Plant and Materials intended for the Works	Determine and certify an amount for Plant and Materials which have been sent to the Site for incorporation in the Permanent Works.	For inclusion in each Interim Payment Certificate.	None.
14.6 Issue of Interim Payment Certificates	a) Issue to the Employer an Interim Payment Certificate. b) In the case that the certified amount would be less than the minimum amount of Interim Payment Certificates stated in the Appendix to Tender, give notice to the Contractor.	a) Within 28 days after receiving a Statement from the Contractor. b) None	If late certification results in the Employer not making payment within the stated period, the Contractor is entitled to receive financing charges (Sub-Clause 14.7).
14.9 Payment of Retention Money	a) Certify the first half of the Retention Money. b) Certify the outstanding balance of the Retention Money.	a) When the Taking-Over Certificate has been issued for the Works. b) Promptly after the latest of the expiry dates of the Defects Notification Periods.	If late certification results in the Employer not making payment within the stated period, the Contractor is entitled to receive financing charges (Sub-Clause 14.7).
14.10 Statement at Completion	Issue to the Employer an Interim Payment Certificate.	Within 28 days after receiving a Statement at completion.	If late certification results in the Employer not making payment within the stated period, the Contractor is entitled to receive financing charges (Sub-Clause 14.7).

THE OBLIGATIONS OF THE ENGINEER (continued)

CLAUSE	OBLIGATIONS	TIME FRAME	SPECIFIC CONSEQUENCES OF NON-COMPLIANCE
14.11 Application for Final Payment Certificate	a) Cooperate with the Contractor to agree and approve the form for the final statement. b) In the case that a dispute exists, deliver to the Employer (with a copy to the Contractor) an Interim Payment Certificate for the agreed parts of the draft final statement.	None.	None.
14.13 Issue of Final Payment Certificate	a) Issue, to the Employer the Final Payment Certificate. b) In the case that the Contractor has not applied for a Final Payment Certificate, request the Contractor to do so. c) In the case that the Contractor fails to submit an application within a period of 28 days, issue the Final Payment Certificate for such amount as the Engineer fairly determines to be due.	a) Within 28 days after receiving the Final Statement and written discharge. b) None. c) None.	If late certification results in the Employer not making payment within the stated period, the Contractor is entitled to receive financing charges (Sub-Clause 14.7).
15 Termination by Employer			
15.3 Valuation at Date of Termination	Agree or determine the value of the Works, Goods, Contractor's Documents and any other sums due to the Contractor for work executed in accordance with the Contract.	As soon as practicable after a notice of termination.	None.
16 Suspension and Termination by Contractor			
16.1 Contractor's Entitlement to Suspend Work	In the case of a notice and claim for Cost or delay for being received, respond to the claim and agree or determine the matters.	Respond within 42 days after receiving a claim or any further particulars supporting a previous claim (Sub-Clause 20.1).	None.

CLAUSE	OBLIGATIONS	TIME FRAME	SPECIFIC CONSEQUENCES OF NON-COMPLIANCE
17 Risk and Responsibility			
17.4 Consequences of Employer's Risks	In the case of a notice and claim for delay or Cost being received, respond to the claim and agree or determine the matters.	Respond within 42 days after receiving a claim or any further particulars supporting a previous claim (Sub-Clause 20.1).	None.
19 Force Majeure			
19.4 Consequences of Force Majeure	In the case of a notice and claim for delay or Cost being received, respond to the claim and agree or determine the matters.	Respond within 42 days after receiving a claim or any further particulars supporting a previous claim (Sub-Clause 20.1).	None.
19.6 Optional Termination, Payment and Release	In the case of termination, determine the value of the work done and issue a Payment Certificate.	Upon termination.	None.
20 Claim, Disputes and Arbitration			
20.1 Contractor's Claims	a) In the case of a claim being received, respond with approval, or with disapproval and detailed comments. b) Agree or determine the extension of the Time for Completion and/or the additional payment.	a) Within 42 days after receiving the claim or any further particulars supporting a previous claim (Sub-Clause 20.1). b) None.	None.

GENERAL CONDITIONS OF DISPUTE ADJUDICATION AGREEMENT

Annex – Procedural Rules			
3.	Attend Site visits by the DAB.	None.	None.

THE OBLIGATIONS OF THE DISPUTE ADJUDICATION BOARD

CLAUSE	OBLIGATIONS	TIME FRAME	SPECIFIC CONSEQUENCES OF NON-COMPLIANCE
GENERAL CONDITIONS			
20 Claims, Disputes and Arbitration			
20.4 Obtaining Dispute Adjudication Board's Decision	Give a decision on any dispute referred to the DAB.	Within 84 days after receiving such reference.	Either Party may commence arbitration.
GENERAL CONDITIONS OF DISPUTE ADJUDICATION AGREEMENT			
3 Warranties	a) Be impartial and independent of the Employer, the Contractor and the Engineer. b) Disclose to the Parties and to the Other Members, any fact or circumstance which might appear inconsistent with his/her warranty and agreement of impartiality and independence.	a) None. b) Promptly.	None.

CLAUSE	OBLIGATIONS	TIME FRAME	SPECIFIC CONSEQUENCES OF NON-COMPLIANCE
4 General Obligations of the Member	a) Have no interest, financial or otherwise in the Parties or the Engineer, nor any financial interest in the Contract. b) Not previously have been employed as a consultant or otherwise by the Parties or the Engineer, except as disclosed in writing. c) Disclose in writing to the Parties and the Other Members, any professional or personal relationships with any director, officer or employee of the Parties or the Engineer and any previous involvement in the overall project of which the Contract forms part. d) Not, for the duration of the Dispute Adjudication Agreement, be employed as a consultant or otherwise by the Parties or the Engineer, except as may be agreed in writing. e) Comply with the procedural rules and with Sub-Clause 20.4 of the Conditions of Contract. f) Not give advice to the Parties, the Employer's Personnel or the Contractor's Personnel concerning the conduct of the Contract, other than in accordance with the procedural rules. g) Not enter into discussions or make any agreement with the Employer, the Contractor or the Engineer regarding employment by any, of them after ceasing to act under the Dispute Adjudication Agreement. h) Ensure his/her availability for all site visits and hearings as are necessary. i) Become conversant with the Contract and with the progress of the Works by studying all documents received. j) Treat the details of the Contract and all the DAB's activities and hearings as private and confidential. k) Be available to give advice and opinions on any matter relevant to the Contract when requested by both of the Parties.	None.	None.

THE OBLIGATIONS OF THE DISPUTE ADJUDICATION BOARD

THE OBLIGATIONS OF THE DISPUTE ADJUDICATION BOARD (continued)

CLAUSE	OBLIGATIONS	TIME FRAME	SPECIFIC CONSEQUENCES OF NON-COMPLIANCE
5 General Obligations of the Employer and the Contractor	a) Not be appointed as an arbitrator in any arbitration under the Contract.	None.	None.
6 Payment	a) Submit invoices for payment of the monthly retainer and air fares. b) Submit invoices for other expenses and for daily fees.	a) Quarterly in advance. b) Following the conclusion of a site visit or hearing.	None.
Annex – Procedural Rules			
1.	Visit the site.	a) At intervals of not more than 140 days. b) At times of critical construction events. c) At the request of either the Employer or the Contractor.	None.
2.	a) Agree the timing of and agenda for each site visit with the Parties. b) In the absence of agreement by the Parties, decide the timing of and agenda for each site visit.	None.	None.
3.	Prepare a report on the DAB's activities during the visit and send copies to the Employer and the Contractor.	At the conclusion of each site visit and before leaving the site.	None.
4.	Copy all communications to the Parties.	None.	None.

CLAUSE	OBLIGATIONS	TIME FRAME	SPECIFIC CONSEQUENCES OF NON-COMPLIANCE
5(a)	a) Act fairly and impartially as between the Parties. b) Give each of the Parties a reasonable opportunity of putting his case and responding to the other's case.	None.	None.
5(b).	Adopt procedures suitable to the dispute, avoiding unnecessary delay or expense.	None.	None.
6.	In the case of a hearing on the dispute, decide on the date and place for the hearing.	None.	None.
9.	a) Not express any opinions during any hearing concerning the merits of any arguments advanced by the Parties. b) Make and give a decision in accordance with Sub-Clause 20.4, or as otherwise agreed by the Employer and the Contractor in writing. c) If the DAB comprises three persons: I. Convene in private after a hearing. II. Endeavour to reach a unanimous decision.	None.	None.

THE OBLIGATIONS OF THE DISPUTE ADJUDICATION BOARD

Chapter 2
The Pink Book

Conditions of Contract for Construction, Multilateral Development Bank Harmonised Edition for Building and Engineering Works Designed by the Employer, June 2010

The FIDIC Contracts: Obligations of the Parties, First Edition. Andy Hewitt.
© 2014 John Wiley & Sons, Ltd. Published 2014 by John Wiley & Sons, Ltd.

THE OBLIGATIONS OF THE EMPLOYER

CLAUSE	OBLIGATIONS	TIME FRAME	SPECIFIC CONSEQUENCES OF NON-COMPLIANCE
GENERAL CONDITIONS			
1 General Provisions			
1.6 Contract Agreement	Enter into a Contract Agreement with the Contractor.	Within 28 days after the Contractor receives the Letter of Acceptance, unless agreed otherwise.	None.
1.8 Care and Supply of Documents	a) Keep the Specification and Drawings in custody and care. b) Supply two copies of the Contract and of each subsequent Drawing to the Contractor. c) Give notice of errors or defects in any document prepared for use in executing the Works.	None.	None.
1.13 Compliance with Laws	Obtain the planning, zoning or similar permission for the Permanent Works, and any other permissions described in the Specification as having been (or being) obtained by the Employer.	None.	None.
2 The Employer			
2.1 Right of Access to the Site	a) Give the Contractor right of access to, and possession of, all parts of the Site. b) Give the Contractor possession of any foundation, structure, plant or means of access if required.	a) Within the time (or times) stated in the Appendix to Tender, or if not stated, to enable the Contractor to proceed in accordance with the programme submitted under Sub-Clause 8.3 [Programme]. b) In the time and manner stated in the Specification.	Contractor shall be entitled to an extension of time and payment of Cost plus reasonable profit.

THE OBLIGATIONS OF THE EMPLOYER (continued)

CLAUSE	OBLIGATIONS	TIME FRAME	SPECIFIC CONSEQUENCES OF NON-COMPLIANCE
2.2 Permits, Licences or Approvals	Provide reasonable assistance to the Contractor at the request of the Contractor: a) By obtaining copies of the Laws of the Country which are relevant but not readily available and b) For the Contractor's applications for any permits, licences or approvals required by the Laws of the Country.	None.	None.
2.3 Employer's Personnel	Ensure that Employer's personnel and Employer's other contractors cooperate with the Contractor and take actions similar to those which the Contractor is required to take under Sub-Clause 4.8 [Safety Procedures] and under Sub-Clause 4.18 [Protection of the Environment].	None.	None.
2.4 Employer's Financial Arrangements	a) Submit reasonable evidence that financial arrangements have been made and are being maintained which will enable the Employer to pay the Contract Price punctually. b) In the case that material changes are made to the financial arrangements, give notice to the Contractor. c) In the case of the Bank suspending disbursements under its loan, give notice to the Contractor. d) Provide reasonable evidence of alternative funding.	a) Within 28 days after receiving any request from the Contractor. b) Before making the changes. c) Within 7 days of the suspension notification. d) None.	Contractor entitled to suspend work, reduce the rate of work and to an extension of time and additional payment as a result of such actions (Sub-Clause 16.1).
2.5 Employer's Claims	Give notice and particulars to the Contractor if the Employer considers himself to be entitled to any payment under any Clause of the Conditions or otherwise in connection with the Contract and/or to any extension of the Defects Notification Period (this obligation may be carried out by the Engineer).	As soon as practicable and no longer than 28 days after becoming aware of the event or circumstances giving rise to the claim.	None.

CLAUSE	OBLIGATIONS	TIME FRAME	SPECIFIC CONSEQUENCES OF NON-COMPLIANCE
3 The Engineer			
3.1 Engineer's Duties and Authority	a) Appoint the Engineer to carry out the duties assigned to him in the Contract. b) Inform the Contractor of any change in the authority of the Engineer.	None.	None.
3.4 Replacement of the Engineer	a) Give notice to the Contractor of the name, address and relevant experience of the intended replacement Engineer. b) In the case that the Contractor raises reasonable objection by notice to the Employer with supporting particulars, give full and fair consideration to the objection.	a) Not less than 21 days before the intended date of replacement. b) None.	None.
4 The Contractor			
4.2 Performance Security	a) Cooperate with the Contractor to agree the entity, country (or other jurisdiction) for the issue of the Performance Security. b) Cooperate with the Contractor to agree the form of Performance Security if not in the form annexed to the Particular Conditions. c) Not to make a claim under the Performance Security, except for amounts to which the Employer is entitled under the Contract. d) Return the Performance Security to the Contractor.	a) None. b) None. c) None. d) Within 21 days after receiving a copy of the Performance Certificate.	None.
4.10 Site Data	a) Make available to the Contractor all relevant data in the Employer's possession on sub-surface and hydrological conditions at the Site, including environmental aspects. b) Make available to the Contractor all such data which come into the Employer's possession after the Base Date.	a) Prior to the Base Date. b) None.	None.

THE OBLIGATIONS OF THE EMPLOYER

Pink Book

THE OBLIGATIONS OF THE EMPLOYER (continued)

CLAUSE	OBLIGATIONS	TIME FRAME	SPECIFIC CONSEQUENCES OF NON-COMPLIANCE
4.13 Rights of Way and Facilities	Provide effective access to and possession of the Site including special and/or temporary rights of way.	None.	None.
4.20 Employer's Equipment and Free-Issue Material	a) Make the Employer's Equipment (if any) available for the use of the Contractor in the execution of the Works in accordance with the details, arrangements and prices stated in the Specification. b) Supply, free of charge, the "free-issue materials" (if any) in accordance with the details stated in the Specification. c) Rectify any notified shortage, defect or default in the free-issue materials.	a) As specified in the Contract. b) As specified in the Contract. c) Immediately.	None.
4.24 Fossils	Take possession and care of fossils, coins, articles of value or antiquity, and structures and other remains, or items of geological or archaeological interest found on the Site.	None.	None.
6 Staff and Labour			
6.12 Foreign Personnel	Use best endeavours to assist the Contractor in connection with bringing in the Contractor's personnel.	In a timely and expeditious manner.	None.
8 Commencement, Delays and Suspension			
8.6 Rate of Progress	Pay additional costs of revised methods, including acceleration measures instructed by the Engineer to reduce delays listed under sub-Clause 8.4.	None.	None.

CLAUSE	OBLIGATIONS	TIME FRAME	SPECIFIC CONSEQUENCES OF NON-COMPLIANCE
10 Employer's Taking Over			
10.1 Taking Over of the Works and Sections	Take over the Works.	When completed in accordance with the Contract and a Taking-Over Certificate has been issued.	None.
10.2 Taking Over of Parts of the Works	Not to use any part of the Works (other than as a temporary measure which is either specified in the Contract or agreed by both Parties) unless and until the Engineer has issued a Taking-Over Certificate for this part.	None.	If the Contractor incurs Cost as a result of the Employer taking over and/or using a part of the Works, the Contractor shall be entitled to payment of any such Cost plus profit.
11 Defects Liability			
11.2 Cost of Remedying Defects	Notify the Contractor (or ensure that notice is given by others) of any work to be remedied if due to any cause outside the provisions of the Contract.	Promptly.	None.
11.4 Failure to Remedy Defects	In the case of failure by the Contractor to remedy any defect or damage, notify the Contractor (or ensure that notice is given by others) of the date by which the defect or damage is to be remedied.	Within reasonable time.	None.

Pink Book

THE OBLIGATIONS OF THE EMPLOYER (continued)

CLAUSE	OBLIGATIONS	TIME FRAME	SPECIFIC CONSEQUENCES OF NON-COMPLIANCE
14 Contract Price and Payment			
14.2 Advance Payment	a) Make the advance payment, as an interest-free loan for mobilisation. b) Cooperate with the Contractor to approve the form of the advance payment guarantee.	a) As stated in the Contract Data and when the Contractor submits a guarantee in accordance with this Sub-Clause. b) None.	a) Contractor entitled to suspend work, reduce the rate of work and to an extension of time and additional payment as a result of such actions (Sub-Clause 16.1) and payment of financing charges to the Contractor (Sub-Clause 14.8). b) None.
14.5 Plant and Material intended for the Works	Cooperate with the Contractor to approve the form of a bank guarantee for shipped Plant and Materials.	None.	None.

CLAUSE	OBLIGATIONS	TIME FRAME	SPECIFIC CONSEQUENCES OF NON-COMPLIANCE
14.7 Payment	a) Pay the first instalment of the advance payment. b) Pay the amount certified in each Interim Payment Certificate. c) Pay the amount certified in the Final Payment Certificate.	a) Within 42 days after issuing the Letter of Acceptance or within 21 days after receiving the documents in accordance with Sub-Clause 4.2 [Performance Security] and Sub-Clause 14.2 [Advance Payment], whichever is later. b) Within 56 days after the Engineer receives the Statement and supporting documents. c) Within 56 days after the Employer receives the Final Payment Certificate.	Contractor entitled to suspend work, reduce the rate of work and to an extension of time and additional payment as a result of such actions (Sub-Clause 16.1) and payment of financing charges to the Contractor Sub-Clause 14.8).
14.9 Payment of Retention Money	a) Pay the Contractor the first half of the Retention Money. b) If necessary, cooperate with the Contractor to approve the form of a bank guarantee for the second half of the Retention Money. c) Pay the Contractor the second half of the Retention Money. d) Return the guarantee to the Contractor.	a) Within 56 days after the Engineer receives the Statement and supporting documents (Sub-Clause 14.7). b) None. c) On receipt of the Engineer's payment certificate. d) Within 21 days of receiving a copy of the Performance certificate.	a) The Contractor is entitled to receive financing charges (Sub-Clause 14.7). b) None. c) The Contractor is entitled to receive financing charges (Sub-Clause 14.7). d) None.
14.15 Currencies of Payment	Pay the Contractor in the currency or currencies named in the Schedule of Payment Currencies.	None.	None.

THE OBLIGATIONS OF THE EMPLOYER (continued)

CLAUSE	OBLIGATIONS	TIME FRAME	SPECIFIC CONSEQUENCES OF NON-COMPLIANCE
15 Termination by Employer			
15.2 Termination by Employer	a) Give notice of intention to terminate the Contract. b) Give notice of release of the Contractor's Equipment and Temporary Works.	a) 14 days prior to the termination date, or b) immediately in the case of the Contractor becoming bankrupt or gives or offers bribes or gratuities (or similar as defined in the clause). b) On completion of the Works.	None.
15.4 Payment after Termination	Pay the balance due to the Contractor after recovering any losses, damages and extra costs.	None.	None.
15.5 Employer's Entitlement to Termination for Convenience	a) Give notice of intention to terminate the Contract. b) Return the Performance Security. c) Not to terminate the Contract in order to execute the Works himself or to arrange for the Works to be executed by another contractor.	a) 28 days prior to the termination date. b) None. c) None.	None.
15.6 Corrupt or Fraudulent Practices	Give notice of intention to terminate the Contract in the case of corrupt, fraudulent, collusive or coercive practice.	14 days prior to termination date.	None.

CLAUSE	OBLIGATIONS	TIME FRAME	SPECIFIC CONSEQUENCES OF NON-COMPLIANCE
16 Suspension and Termination by Contractor			
16.4 Payment on Termination	a) Return the Performance Security to the Contractor. b) Pay the Contractor in accordance with Sub-Clause 19.6 *[Optional Termination, Payment and Release]*. c) Pay to the Contractor the amount of any loss of profit or other loss or damage sustained by the Contractor as a result of this termination.	Promptly.	None.
17 Risk and Responsibility			
17.1 Indemnities	Indemnify and hold harmless the Contractor, the Contractor's Personnel and their respective agents, against and from all claims, damages, losses and expenses in respect of bodily injury, disease or death which is attributable to any negligence, wilful act or breach of the Contract by the Employer, the Employer's Personnel or agents.	None.	None.
17.5 Intellectual and Industrial Property Rights	a) Give notice of any claim under this clause. b) Indemnify and hold the Contractor harmless against and from any claim, which is or was an unavoidable result of the Contractor's compliance with the Contract or as a result of any Works being used by the Employer. c) If requested by the Contractor, assist in contesting a claim. d) Not to make any admission, which might be prejudicial to the Contractor.	a) Within 28 days of receiving a claim. b) None. c) None. d) None.	a) Waiver of right to indemnity. b) None. c) None. d) None.

Pink Book

THE OBLIGATIONS OF THE EMPLOYER (continued)

CLAUSE	OBLIGATIONS	TIME FRAME	SPECIFIC CONSEQUENCES OF NON-COMPLIANCE
18 Insurance			
18.1 General Requirements for Insurances	a) Cooperate with the Contractor to approve the terms of insurances. b) Effect and maintain the insurances in terms consistent with any terms agreed between the Parties, wherever the Employer is the insuring Party. c) Submit evidence that the insurance has been effected, provide copies of the policies and submit evidence of payment. d) Inform the insurers of any relevant changes to the execution of the Works and ensure that insurance is maintained. e) Not to make any material alteration to the terms of any insurance without approval of the Contractor.	a) None. b) None. c) Within the time frames stipulated in the Contract Data. d) As appropriate. e) None.	a) None. b) None. c) The Contractor may effect the insurance and the Contract Price shall be adjusted. d) None. e) None.
19 Force Majeure			
19.2 Notice of Force Majeure	Give notice to the Contractor in the case that the Employer is, or will be prevented from performing the Employer's obligations by Force Majeure.	Within 14 days of becoming aware of the event.	None.
19.3 Duty to Minimise Delay	a) Use all reasonable endeavours to minimise any delay in the performance of the Contract as a result of Force Majeure. b) Give notice when the effects of the Force Majeure cease.	None.	None.

CLAUSE	OBLIGATIONS	TIME FRAME	SPECIFIC CONSEQUENCES OF NON-COMPLIANCE
19.6 Optional Termination, Payment and Release	In the case of termination due to Force Majeure, pay the Contractor for work carried out and other costs listed in this sub-clause.	Upon termination.	None.
20 Claims, Disputes and Arbitration			
20.2 Appointment of the Dispute Board	a) Jointly appoint the Dispute Board (DB). b) Not to consult the DB without the agreement of the Contractor. c) Not to act alone in the termination of any member of the DB.	a) By the date stated in the Contract Data. b) None. c) None.	a) The appointing entity or official named in the Appendix to Tender shall appoint (Sub-Clause 20.3). d) None. e) None.
20.4 Obtaining Dispute Board's Decision	a) Make available to the DB additional information, access to the Site, and appropriate facilities as the DB may require. b) Give effect to a DB decision unless and until it is revised in an amicable settlement or an arbitral award.	a) Promptly. b) None.	c) None. a) The matter may be referred to arbitration (Sub-Clause 20.7).
20.5 Amicable Settlement	In the case of a Notice of Dissatisfaction being issued, attempt to settle the dispute amicably.	Within 56 days of the notice.	Arbitration may be commenced.

Pink Book

THE OBLIGATIONS OF THE EMPLOYER (continued)

CLAUSE	OBLIGATIONS	TIME FRAME	SPECIFIC CONSEQUENCES OF NON-COMPLIANCE
GENERAL CONDITIONS OF DISPUTE ADJUDICATION AGREEMENT			
5 General Obligations of the Employer and the Contractor	a) Not to request advice from, or consultation with a DB Member regarding the Contract otherwise than in the normal course of the DB's activities. b) In the case of the DB Member being required to make a site visit or attend a hearing, provide appropriate security for a sum equivalent to the reasonable expenses to be incurred by the Member (this may be undertaken by the Contractor).	None.	None.
6 Payment	a) Pay one half of the DB fees to the Contractor. b) In the Case of the Contractor failing to pay the DAB Member, pay the due fees.	a) Within the monthly payments to the Contractor. b) None.	The DB Member may suspend services or resign the appointment.
Annex – Procedural Rules			
2.	Jointly agree the timing of and agenda for each Site visit by the DB	None.	Timing and agenda shall be decided by the DB.
3.	a) Attend site visits by the DB. b) Co-ordinate site visits by the DB in co-operation with the Contractor. c) Ensure the provision of appropriate conference facilities and secretarial and copying services to the DB.	None.	None.
4.	a) Furnish to each DB member one copy of all documents, which the DB may request. b) Copy the Contractor on all communications between the DB and the Employer.	None.	None.

THE OBLIGATIONS OF THE CONTRACTOR

CLAUSE	OBLIGATIONS	TIME FRAME	SPECIFIC CONSEQUENCES OF NON-COMPLIANCE
GENERAL CONDITIONS			
1 Definitions			
1.6 Contract Agreement	Enter into a Contract Agreement.	Within 28 days after the Contractor receives the Letter of Acceptance, unless the Particular Conditions establish otherwise.	None.
1.8 Care and Supply of Documents	a) Supply to the Engineer six copies of each of the Contractor's Documents. b) Keep, on the Site, a copy of the Contract, publications named in the Specification, the Contractor's Documents (if any), the Drawings and Variations and other communications given under the Contract. c) In the case of an error in a document, give notice to the other Party.	a) None. b) None. c) Promptly.	None.
1.9 Delayed Drawings or Instructions	a) Give notice to the Engineer whenever the Works are likely to be delayed or disrupted if any necessary drawing or instruction is not issued to the Contractor within a particular time. b) Give a further notice to the Engineer if the Contractor suffers delay and/or incurs Cost.	a) Within a reasonable time. b) As soon as practicable and not later than 28 days after the Contractor became aware, or should have become aware of the event or circumstance (Sub-Clause 20.1).	a) None. b) Loss of entitlement to an extension to the Time for Completion and additional payment (Sub-Clause 20.1).
1.12 Confidential Details	Disclose all such confidential and other information as the Engineer may reasonably be required in order to verify compliance with the Contract.	None.	None.

Pink Book

THE OBLIGATIONS OF THE CONTRACTOR (continued)

CLAUSE	OBLIGATIONS	TIME FRAME	SPECIFIC CONSEQUENCES OF NON-COMPLIANCE
1.13 Compliance with Laws	a) Comply with applicable Laws. b) Give all notices, pay all taxes, duties and fees, and obtain all permits, licences and approvals, as required by the Laws.	None.	None.
1.14 Joint and Several Liability	a) In the case of a joint venture, consortium or other unincorporated grouping of two or more persons, notify the Employer of the leader. b) Not to alter the composition or legal status of the joint venture without the prior consent of the Employer.	None.	None.
1.15 Inspections and Audit by the Bank	Permit the bank and/or persons appointed by the bank to inspect the Site and/or the Contractor's accounts and records and to have the same audited by the Bank.	None.	None.
2 The Employer			
2.1 Right of Access to the Site	Give notice to the Engineer if the Contractor suffers delay and/or incurs Cost as a result of failure to give right of access and possession of the Site.	As soon as practicable and not later than 28 days after the Contractor became aware, or should have become aware of the event or circumstance (Sub-Clause 20.1).	Loss of entitlement to an extension to the Time for Completion and additional payment (Sub-Clause 20.1).
3 The Engineer			
3.3 Instructions of the Engineer	a) Only take instructions from the Engineer, or from an assistant to whom the appropriate authority has been delegated. b) Comply with the instructions given by the Engineer or delegated assistant on any matter related to the Contract.	None.	None.

CLAUSE	OBLIGATIONS	TIME FRAME	SPECIFIC CONSEQUENCES OF NON-COMPLIANCE
4 The Contractor			
4.1 Contractor's General Obligations	a) Design (to the extent specified in the Contract), execute and complete the Works in accordance with the Contract and with the Engineer's instructions and remedy any defects in the Works. b) Provide the required Plant and Contractor's Documents specified in the Contract and all Contractor's Personnel, Goods, consumables and other things and services whether of a temporary or permanent nature. c) Be responsible for the adequacy, stability and safety of all Site operations and of all methods of construction. d) Be responsible for all Contractor's Documents, Temporary Works, and such design of each item of Plant and Materials as is required for the item to be in accordance with the Contract. e) Submit details of the arrangements and methods proposed for the execution of the Works. f) Design any part of the Permanent Works if specified in the Contract. g) Submit to the Engineer the "as-built" documents and operation and maintenance manuals prior to the Tests on Completion.	None.	None.

THE OBLIGATIONS OF THE CONTRACTOR (continued)

CLAUSE	OBLIGATIONS	TIME FRAME	SPECIFIC CONSEQUENCES OF NON-COMPLIANCE
4.2 Performance Security	a) Obtain a Performance Security for proper performance and deliver to the Employer. b) Ensure that the Performance Security is valid and enforceable until the Contractor has executed and completed the Works and remedied any defects. c) Extend the validity of the Performance Security until the Works have been completed and any defects have been remedied.	a) Within 28 days after receiving the Letter of Acceptance. b) None. c) As required.	None.
4.3 Contractor's Representative	a) Appoint the Contractor's Representative and give him all authority necessary to act on the Contractor's behalf under the Contract. b) Submit to the Engineer for consent, the name and particulars of the person the Contractor proposes to appoint. c) If consent is withheld or subsequently revoked, or if the appointed person fails to act, submit the name and particulars of another suitable person for such appointment. d) Not, without the prior consent of the Engineer, revoke the appointment of the Contractor's Representative or appoint a replacement. e) If the Contractor's Representative is to be temporarily absent from the Site, appoint a suitable replacement and notify the Engineer.	a) None. b) Prior to the Commencement Date. c) None. d) None. e) None.	None.

CLAUSE	OBLIGATIONS	TIME FRAME	SPECIFIC CONSEQUENCES OF NON-COMPLIANCE
4.4 Subcontractors	a) Not to subcontract the whole of the Works. b) Be responsible for the acts or defaults of any Subcontractor, his agents or employees. c) Obtain prior consent of the Engineer to proposed Subcontractors not named in the Contract. d) Give the Engineer notice of the intended date of the commencement of each Subcontractor's work and of the commencement of such work on the Site. e) Ensure that the requirements imposed upon the Contractor by Sub-Clause 1.12 (*Confidential Details*) apply to subcontractors. f) Where practicable, give opportunity for contractors from the Country to be appointed as Subcontractors.	a) None. b) None. c) None. d) Not less than 28 days. e) None f) None	None.
4.5 Assignment of Benefit of Subcontract	If instructed, assign the benefit of subcontract obligations that extend beyond the Defects Notification Period to the Employer.	None.	None.
4.6 Co-operation	a) Allow appropriate opportunities for carrying out work to the Employer's Personnel, any other contractors employed by the Employer and the personnel of any legally constituted public authorities. b) Submit such documents, which require the Employer to give to the Contractor possession of any foundation, structure, plant or means of access.	a) None. b) In the time and manner stated in the Specification.	None.

THE OBLIGATIONS OF THE CONTRACTOR

Pink Book

THE OBLIGATIONS OF THE CONTRACTOR (continued)

CLAUSE	OBLIGATIONS	TIME FRAME	SPECIFIC CONSEQUENCES OF NON-COMPLIANCE
4.7 Setting Out	a) Set out the Works in relation to original points, lines and levels of reference specified in the Contract or notified by the Engineer. b) Give notice to the Engineer if the Contractor suffers delay and/or incurs Cost as a result of error in the items of reference.	a) None. b) As soon as practicable and not later than 28 days after the Contractor became aware, or should have become aware of the event or circumstance (Sub-Clause 20.1).	a) None. b) Loss of entitlement to an extension to the Time for Completion and additional payment (Sub-Clause 20.1).
4.8 Safety Procedures	a) Comply with all applicable safety regulations. b) Take care for the safety of all persons entitled to be on the Site. c) Use reasonable efforts to keep the Site and Works clear of unnecessary obstruction. d) Provide fencing, lighting, guarding and watching of the Works. e) Provide any Temporary Works, which may be necessary for the use and protection of the public and of owners and occupiers of adjacent land.	None.	None.
4.9 Quality Assurance	Institute a quality assurance system and submit details to the Engineer.	Before each design and execution stage has commenced.	None.

CLAUSE	OBLIGATIONS	TIME FRAME	SPECIFIC CONSEQUENCES OF NON-COMPLIANCE
4.12 Unforeseeable Physical Conditions	a) Give notice of adverse physical conditions. b) Continue executing the Works, using such proper and reasonable measures as are appropriate for the physical conditions. c) Comply with any instructions, which the Engineer may give. d) Give a further notice if the Contractor suffers delay and/or incurs Cost due to unforeseen physical conditions.	a) As soon as practicable. b) None. c) None. d) As soon as practicable and not later than 28 days after the Contractor became aware, or should have become aware of the event or circumstance (Sub-Clause 20.1).	a) None. b) None. c) None. d) Loss of entitlement to an extension to the Time for Completion and additional payment (Sub-Clause 20.1).
4.13 Rights of Way and Facilities	Obtain any additional rights of way facilities outside the Site which the Contractor may require for the purposes of the Works.	None.	None.
4.14 Avoidance of Interference	Not to interfere with the convenience of the public, or the access to and use and occupation of all roads and footpaths.	None.	None.
4.15 Access Route	a) Use reasonable efforts to prevent any road or bridge from being damaged. b) Be responsible for any maintenance, which may be required for the use of access routes. c) Provide all necessary signs or directions along access routes. d) Obtain any permission, which may be required from the relevant authorities for use of routes, signs and directions.	None.	None.

THE OBLIGATIONS OF THE CONTRACTOR (continued)

CLAUSE	OBLIGATIONS	TIME FRAME	SPECIFIC CONSEQUENCES OF NON-COMPLIANCE
4.16 Transport of Goods	a) Give the Engineer notice of the date on which any Plant or a major item of other Goods will be delivered to the Site. b) Be responsible for packing, loading, transporting, receiving, unloading, storing and protecting all Goods and other things required for the Works.	a) 21 days before delivery. b) None.	None.
4.17 Contractor's Equipment	a) Be responsible for all Contractor's Equipment. b) Not to remove from the Site any major items of Contractor's Equipment without the consent of the Engineer.	None.	None.
4.18 Protection of the Environment	a) Take all reasonable steps to protect the environment and to limit damage and nuisance to people and property. b) Ensure that emissions, surface discharges and effluent shall not exceed the values indicated in the Specification and shall not exceed the values prescribed by applicable Laws.	None.	None.
4.19 Electricity, Water and Gas	a) Be responsible for the provision of all power, water and other services. b) Provide any apparatus necessary for use of services as may be available on the Site and for measuring the quantities consumed. c) Pay the Employer for the use of services available on the Site.	None.	None.

CLAUSE	OBLIGATIONS	TIME FRAME	SPECIFIC CONSEQUENCES OF NON-COMPLIANCE
4.20 Employer's Equipment and Free-Issue Materials	a) Be responsible for the Employers' Equipment when used by the Contractor. b) Pay the Employer for the use of the Employers' Equipment. c) Inspect free-issue materials. d) Give notice of any shortage, defect or default in the free-issue materials.	a) None. b) None. c) None. d) Promptly.	None.
4.21 Progress Reports	Prepare and submit monthly progress reports.	Monthly, within 7 days of the period to which the report relates.	None.
4.22 Security of the Site	Keep unauthorised persons off the Site.	None.	None.
4.23 Contractor's Operations on Site	a) Confine the operations to the Site and to any additional areas agreed as working areas. b) Take all necessary precautions to keep the Contractor's Equipment and Contractor's Personnel within the Site and any agreed working areas. c) Keep the Site free from all unnecessary obstruction. d) Store or dispose of any Contractor's Equipment or surplus materials. e) Clear away and remove from the Site any wreckage, rubbish and Temporary Works. f) Leave the Site and the Works in a clean and safe condition.	None.	None.

THE OBLIGATIONS OF THE CONTRACTOR (continued)

CLAUSE	OBLIGATIONS	TIME FRAME	SPECIFIC CONSEQUENCES OF NON-COMPLIANCE
4.24 Fossils	a) Take reasonable precautions to prevent the Contractor's Personnel or other persons from removing or damaging fossils, coins, articles of value or antiquity, structures and other remains or items of geological or archaeological interest. b) Give notice of the finding of such items. c) Give further notice if the Contractor suffers delay and/or incurs Cost as a result of such items.	a) None. b) Upon discovery. c) As soon as practicable and not later than 28 days after the Contractor became aware, or should have become aware of the event or circumstance (Sub-Clause 20.1).	a) None. b) None. c) Loss of entitlement to an extension to the Time for Completion and additional payment (Sub-Clause 20.1).
5 Nominated Subcontractors			
5.3 Payments to Nominated Subcontractors	Pay to the nominated Subcontractor the amounts which the Engineer certifies.	None.	None.
6 Staff and Labour			
6.1 Engagement of Staff and Labour	Make arrangements for the engagement of all staff and labour, local or otherwise, and for their payment, housing, feeding and transport.	None.	None.
6.2 Rates of Wages and Conditions of Labour	Pay rates of wages and observe conditions of labour which are not lower than those established for the trade or industry where the work is carried out.	None.	None.

CLAUSE	OBLIGATIONS	TIME FRAME	SPECIFIC CONSEQUENCES OF NON-COMPLIANCE
6.3 Persons in the Service of the Employer	Not to recruit, or attempt to recruit, staff and labour from amongst the Employer's personnel.	None.	None.
6.4 Labour Laws	a) Comply with all the relevant labour Laws. b) Require employees to obey all applicable Laws.	None.	None.
6.5 Working Hours	Obtain the consent of the Engineer if working outside the normal working hours.	None.	None.
6.6 Facilities for Staff and Labour	a) Provide and maintain all necessary accommodation and welfare facilities for the Contractor's Personnel. b) Provide facilities for the Employer's Personnel as stated in the Specification. c) Not to permit any of the Contractor's Personnel to maintain any temporary or permanent living quarters within the structures forming part of the Permanent Works.	None.	None.

Pink Book

THE OBLIGATIONS OF THE CONTRACTOR (continued)

CLAUSE	OBLIGATIONS	TIME FRAME	SPECIFIC CONSEQUENCES OF NON-COMPLIANCE
6.7 Health and Safety	a) Take all reasonable precautions to maintain the health and safety of the Contractor's Personnel.	a) None.	None.
	b) Ensure that medical staff, first aid facilities, sick bay and ambulance service are available at all times and that suitable arrangements are made for all necessary welfare and hygiene requirements and for the prevention of epidemics.	b) None.	
	c) Appoint an accident prevention officer and whatever is required by this person to exercise this responsibility and authority.	c) None.	
	d) Send to the Engineer details of any accident.	d) As soon as practicable after its occurrence.	
	e) Maintain records and make reports concerning health, safety, welfare and damage to property.	e) None.	
	f) Conduct an HIV awareness programme and undertake other measure as specified in the Contract with regard to HIV.	f) None.	
	g) Conduct information, education and consultation, communication campaigns with regard to Sexually Transmitted Diseases (STD), Sexually Transmitted Infections (STI) and HIV/AIDS.	g) None.	
	h) Provide condoms for all Site staff and labour.	h) None.	
	i) Provide for STI and HIV/AIDS screening, diagnosis, counselling and referral.	i) None.	
	j) Include in the programme an alleviation programme in respect of STDs, STIs and HIV/AIDS.	j) None.	

CLAUSE	OBLIGATIONS	TIME FRAME	SPECIFIC CONSEQUENCES OF NON-COMPLIANCE
6.8 Contractor's Superintendence	Provide all necessary superintendence to plan, arrange, direct, manage, inspect and test the work.	None.	None.
6.10 Records of Contractor's Personnel and Equipment	Submit to the Engineer, details showing the number of each class of Contractor's Personnel and of each type of Contractor's Equipment on the Site.	Each calendar month.	None.
6.11 Disorderly Conduct	Take all reasonable precautions to prevent any unlawful, riotous or disorderly conduct by, or amongst the Contractor's Personnel.	None.	None.
6.12 Foreign Personnel	a) In the case of foreign personnel, ensure that the personnel are provided with the required visas and work permits. b) Be responsible for the return of foreign personnel and in the event of death, make the appropriate arrangements for their return or burial.	None.	None.
6.13 Supply of Foodstuffs	Arrange for the provision of suitable food for the Contractor's personnel at reasonable prices.	None.	None.
6.14 Supply of Water	Provide an adequate supply of water.	None.	None.
6.15 Measures Against Insect and Pest Nuisance	a) Take necessary precautions to protect the Contractor's personnel from insect and pest nuisance, and to reduce danger to their health. b) Comply with regulations of the local health authorities.	None.	None.
6.16 Alcoholic Liquor or Drugs	Not to sell, give, barter or otherwise dispose of any alcoholic liquor or drugs, or permit the same by the Contractor's Personnel.	None.	None.

THE OBLIGATIONS OF THE CONTRACTOR

Pink Book

THE OBLIGATIONS OF THE CONTRACTOR (continued)

CLAUSE	OBLIGATIONS	TIME FRAME	SPECIFIC CONSEQUENCES OF NON-COMPLIANCE
6.17 Arms and Ammunition	Not to sell, give, barter, or otherwise dispose of any arms or ammunition or permit the same by the Contractor's Personnel.	None.	None.
6.18 Festivals and Religious Customs	Respect the Country's festivals, days of rest or other customs.	None.	None.
6.19 Funeral Arrangements	Be responsible for making funeral arrangement for any employee who may die while engaged upon the Works.	None.	None.
6.20 Prohibition of Forced or Compulsory Labour	Not to employ forced or compulsory labour.	None.	None.
6.21 Prohibition of Harmful Child Labour	Not to employ any child to perform any work that is economically exploitive or is likely to be hazardous to, or interfere with, the child's education or be harmful to the child's health or physical, mental spiritual, moral or social development.	None.	None.
6.22 Employment Records of Workers	a) Keep records of the employment of labour at the Site. b) Summarise the records and submit them to the Engineer.	a) None. b) On a monthly basis.	None.

CLAUSE	OBLIGATIONS	TIME FRAME	SPECIFIC CONSEQUENCES OF NON-COMPLIANCE
6.23 Workers' Organisations	a) Comply with laws that recognise workers' rights to form and join workers organisations. b) In the case of laws that restrict workers' organisations, enable alternative means for the Contractor's Personnel to express grievances and protect their rights. c) Not discourage or discriminate against the Contractor's Personnel from forming or joining workers' organisations. d) Engage with the representatives of workers' organisations.	None.	None.
6.24 Non-Discrimination and Equal Opportunity	a) Base the employment relationship on the basis of equal opportunity and fair treatment. b) Not to discriminate with respect to aspects of the employment relationship	None.	None.
7 Plant, Materials and Workmanship			
7.1 Manner of Execution	Carry out the manufacture of Plant, the production and manufacture of Materials, and all other execution of the Works.	None.	None.
7.2 Samples	Submit samples of Materials and relevant information to the Engineer for consent.	Prior to using the Materials.	None.
7.3 Inspection	a) Give the Employer's Personnel the full opportunity to carry out inspections. b) Give notice to the Engineer to inspect.	a) None. b) Whenever any work is ready and before it is covered up, put out of sight or packaged for storage or transport.	a) None. b) Uncover the work, reinstate and make good at the Contractor's cost.

Pink Book

THE OBLIGATIONS OF THE CONTRACTOR (continued)

CLAUSE	OBLIGATIONS	TIME FRAME	SPECIFIC CONSEQUENCES OF NON-COMPLIANCE
7.4 Testing	a) Provide everything necessary to carry out the specified tests. b) Agree with the Engineer, the time and place for the testing. c) Give notice if the Contractor suffers delay and/or incurs Cost as a result of complying with instructions or a delay for which the Employer is responsible. d) Forward to the Engineer certified reports of the tests.	a) None. b) None. c) As soon as practicable and not later than 28 days after the Contractor became aware, or should have become aware of the event or circumstance (Sub-Clause 20.1). d) Promptly.	a) None. b) None. c) Loss of entitlement to an extension to the Time for Completion and additional payment (Sub-Clause 20.1). d) None.
7.5 Rejection	Make good defects notified by the Engineer.	Promptly.	None.
7.6 Remedial Work	Comply with the instructions of the Engineer with regard to remedial work.	Within a reasonable time as specified in the instruction or immediately if urgency is specified.	The Contractor shall pay costs incurred by the Employer in engaging other persons to carry out the work.
7.8 Royalties	Pay all royalties, rents and other payments for natural Materials obtained from outside the Site and disposal of surplus materials.	None.	None.
8 Commencement, Delays and Suspension			
8.1 Commencement of Works	Commence the execution of the Works and proceed with the Works with due expedition and without delay.	As soon as is reasonably practicable after the Commencement Date.	None.
8.2 Time for Completion	Complete the whole of the Works and each Section within the times specified in the Contract.	None.	Contractor shall pay delay damages to the Employer (Sub-Clause 8.7).

CLAUSE	OBLIGATIONS	TIME FRAME	SPECIFIC CONSEQUENCES OF NON-COMPLIANCE
8.3 Programme	a) Submit a detailed time programme. b) Submit a revised programme. c) Proceed in accordance with the programme. d) Give notice to the Engineer of specific probable future events or circumstances, which may adversely affect the work, increase the Contract Price or delay the execution of the Works. e) Submit a revised programme on receiving a notice from the Engineer that a programme fails to comply with the Contract or to be consistent with actual progress.	a) Within 28 days after receiving the notice of commencement. b) Whenever the previous programme is inconsistent with actual progress or with the Contractor's obligations. c) None. d) Promptly. e) None.	None.
8.4 Extension of Time for Completion	Give notice to the Engineer if the Contractor considers himself to be entitled to an extension of the Time for Completion.	As soon as practicable and not later than 28 days after the Contractor became aware, or should have become aware of the event or circumstance (Sub-Clause 20.1).	Loss of entitlement to an extension to the Time for Completion (Sub-Clause 20.1).
8.6 Rate of Progress	Adopt revised methods in order to expedite progress and complete within the Time for Completion.	None.	None.
8.7 Delay Damages	Pay delay damages in the case of failure to comply with the Time for Completion.	None.	None.
8.8 Suspension of Work	Protect, store and secure such part or the Works in the case of an instruction to suspend the Works.	None.	None.
8.9 Consequences of Suspension	Give notice to the Engineer if the Contractor suffers delay and/or incurs cost as a result of complying with the Engineer's instructions under sub-Clause 8.8.	As soon as practicable and not later than 28 days after the Contractor became aware, or should have become aware of the event or circumstance (Sub-Clause 20.1).	Loss of entitlement to an extension to the Time for Completion and additional payment (Sub-Clause 20.1).

THE OBLIGATIONS OF THE CONTRACTOR

Pink Book

THE OBLIGATIONS OF THE CONTRACTOR (continued)

CLAUSE	OBLIGATIONS	TIME FRAME	SPECIFIC CONSEQUENCES OF NON-COMPLIANCE
8.12 Resumption of Work	a) Jointly examine the Works and the Plant and Materials affected by the suspension, with the Engineer. b) Make good any deterioration, defect or loss.	None.	None.
9 Tests on Completion			
9.1 Contractor's Obligations	a) Carry out the Tests on Completion. b) Give to the Engineer notice of the date after which the Contractor will be ready to carry out each of the Tests on Completion. c) Submit a certified report of the results of these Tests to the Engineer.	a) After providing the documents in accordance with Sub-Clause 4.1(d). b) Not less than 21 days. c) As soon as the Works or a Section have passed the Tests on Completion.	None.
9.2 Delayed Tests	Carry out the Tests if the Engineer gives notice of undue delay.	Within 21 days of the Engineer's notice.	The Employer's Personnel may proceed with the tests at the Contractor's cost.
10 Employer's Taking Over			
10.2 Taking Over of Parts of the Works	a) Carry out any outstanding Tests on Completion. b) Give notice of Costs incurred as a result of the Employer taking over and/or using a part of the Works.	a) As soon as practicable. b) As soon as practicable and not later than 28 days after the Contractor became aware, or should have become aware of the event or circumstance (Sub-Clause 20.1).	a) None. b) Loss of entitlement to additional payment (Sub-Clause 20.1).

CLAUSE	OBLIGATIONS	TIME FRAME	SPECIFIC CONSEQUENCES OF NON-COMPLIANCE
10.3 Interference with Tests on Completion	a) In the case of prevention to the carrying out of the tests, carry out any outstanding Tests on Completion. b) Give notice if the Contractor suffers delay and/or incurs Cost as a result of interference with Tests on Completion.	a) As soon as practicable. b) As soon as practicable and not later than 28 days after the Contractor became aware, or should have become aware of the event or circumstance (Sub-Clause 20.1).	a) None. b) Loss of entitlement to an extension to the Time for Completion and additional payment (Sub-Clause 20.1).
11 Defects Liability			
11.1 Completion of Outstanding Work and Remedying Defects	a) Complete any work which is outstanding on the date stated in a Taking-Over Certificate. b) Execute all work required to remedy defects or damage.	a) Within such reasonable time as is instructed by the Engineer. b) On or before the expiry date of the Defects Notification Period.	a) The Employer may carry out the work himself at the Contractor's cost (Sub-Clause 11.4). b) A reduction in the Contract Price may be made (Sub-Clause 11.4).
11.8 Contractor to Search	If required by the Engineer, search for the cause of any defect.	None.	None.
11.11 Clearance of Site	Remove any remaining Contractor's Equipment, surplus material, wreckage, rubbish and Temporary Works from the Site.	Within 28 days of receipt of the Performance Certificate.	a) The Employer may sell or otherwise dispose of any remaining items. b) Employer entitled to recover costs of disposal.

THE OBLIGATIONS OF THE CONTRACTOR

Pink Book

THE OBLIGATIONS OF THE CONTRACTOR (continued)

CLAUSE	OBLIGATIONS	TIME FRAME	SPECIFIC CONSEQUENCES OF NON-COMPLIANCE
12 Measurement and Evaluation			
12.1 Works to be Measured	a) Show in each payment application, the quantities and other particulars, detailing the amounts to which he considers to be entitled under the Contract. b) Assist the Engineer in making the measurement. c) Supply any particulars requested by the Engineer. d) Examine and agree the records with the Engineer and sign the same when agreed. e) In the case of disagreement with the records, give notice to the Engineer.	a) With each payment application. b) Promptly. c) None. d) As and when requested. e) Within 14 days after examination.	a) None. b,c&d) The Engineer's measurement shall be accepted as accurate. e) None.
12.4 Omissions	In the case where the Contractor will incur cost or not be adequately compensated for omitted work, give notice to the Engineer with supporting particulars.	None.	None.
13 Variations and Adjustments			
13.1 Right to Vary	a) Execute and be bound by each Variation. b) Not make any alteration and/or modification of the Permanent Works, unless and until the Engineer instructs or approves a Variation.	None.	None.
13.3 Variation Procedure	a) Respond in writing to a request for a proposal. b) Not to delay any work whilst awaiting a response. c) Acknowledge receipt of Variation instructions.	a) As soon as practicable. b) None. c) None.	None.

CLAUSE	OBLIGATIONS	TIME FRAME	SPECIFIC CONSEQUENCES OF NON-COMPLIANCE
13.5 Provisional Sums	Produce quotations, invoices, vouchers and accounts or receipts in substantiation of the amounts paid to nominated Subcontractors.	When required by the Engineer.	None.
13.6 Daywork	a) Submit quotations to the Engineer. b) Submit invoices, vouchers and accounts or receipts for Goods. c) Deliver to the Engineer statements, which include the details of the resources used in executing the previous day's work. d) Submit priced statements of these resources.	a) Before ordering Goods for the work to be executed on a Daywork basis. b) When applying for payment. c) Each day. d) Prior to their inclusion in the next Statement under Sub-Clause 14.3.	None.
13.7 Adjustments for Changes in Legislation	Give notice if the Contractor suffers delay and/or incurs Cost as a result of changes in legislation.	As soon as practicable and not later than 28 days after the Contractor became aware, or should have become aware of the event or circumstance (Sub-Clause 20.1).	Loss of entitlement to an extension to the Time for Completion and additional payment (Sub-Clause 20.1).
14 Contract Price and Payment			
14.1 The Contract Price	a) Pay all taxes, duties and fees to be paid under the Contract. b) Submit to the Engineer, a proposed breakdown of each lump sum price in the Schedules.	a) None. b) Within 28 days after the Commencement Date.	None.
14.2 Advance Payment	a) Submit an advance payment guarantee. b) Extend the validity of the guarantee until the advance payment has been repaid.	None.	a) Employer not obliged to make the advance payment. b) None.

Pink Book

THE OBLIGATIONS OF THE CONTRACTOR (continued)

CLAUSE	OBLIGATIONS	TIME FRAME	SPECIFIC CONSEQUENCES OF NON-COMPLIANCE
14.3 Application for Interim Payment Certificates	Submit a Statement in six copies, showing in detail the amounts to which the Contractor considers himself to be entitled.	After the end of each month.	No obligation on the Engineer to certify payment (Sub-Clause 14.6).
14.4 Schedule of Payments	In the case that the Contract does not include a schedule of payments, submit non-binding estimates of the payments expected to become due.	a) First estimate within 42 days after the Commencement Date. b) Revised estimates at quarterly intervals.	None.
14.10 Statement at Completion	Submit a Statement at completion.	Within 84 days after receiving the Taking-Over Certificate for the Works.	None.
14.11 Application for Fina Payment Certificate	a) Submit a draft final statement. b) Submit such further information as the Engineer may reasonably require. c) Prepare and submit the final statement as agreed with the Engineer.	a) Within 56 days after receiving the Performance Certificate. b) Within 28 days of receipt of draft statement. c) None.	a) None. b) None. c) The Engineer will certify an amount as he fairly determines to be due.
14.12 Discharge	Submit a written discharge.	When submitting the Final Statement.	None.
15 Termination by Employer			
15.2 Termination by Employer	a) In the case of a notice of termination being served, leave the Site and deliver any required Goods, Contractor's Documents and other design documents to the Engineer. b) Use best efforts to comply with any reasonable instructions included in the notice. c) Arrange for the removal of Equipment and Temporary Works.	a) None. b) Immediately. c) Promptly.	a) None. b) None. c) Items may be sold by the Employer.

CLAUSE	OBLIGATIONS	TIME FRAME	SPECIFIC CONSEQUENCES OF NON-COMPLIANCE
15.5 Employer's Entitlement to Termination for Convenience	In the case of a notice of termination, cease all further work, hand over Contractor's Documents, Plant, Materials and other work and remove all other Goods from the Site (Sub-Clause 16.3).	28 days from the Employer's notice or return of the Performance Security, whichever is the later.	None.
15.6 Corrupt or Fraudulent Practices	Remove any employee who has been determined to have engaged in corrupt or fraudulent practices and if appropriate appoint a suitable replacement person.	None.	None.
16 Suspension and Termination by Contractor			
16.1 Contractor's Entitlement to Suspend Work	a) Give notice if the Contractor intends to suspend work or reduce the rate of work. b) Resume normal working when the Employer's obligations have been met. c) Give further notice if the Contractor suffers delay and/or incurs Cost as a result of suspending work or reducing the rate of work.	a) 21 days before the intended suspension or reduction in the rate of work. b) As soon as is reasonably practicable. c) As soon as practicable and not later than 28 days after the Contractor became aware, or should have become aware of the event or circumstance (Sub-Clause 20.1).	a) None. b) None. c) Loss of entitlement to an extension to the Time for Completion and additional payment (Sub-Clause 20.1).
16.2 Termination by Contractor	Give notice of intention to terminate.	14 days before the intended termination date.	None.
16.3 Cessation of Work and Removal of Contractor's Equipment	In the case of a notice of termination, cease all further work, hand over Contractor's Documents, Plant, Materials and other work and remove all other Goods from the Site and leave the Site.	After the notice has taken effect.	None.

THE OBLIGATIONS OF THE CONTRACTOR

Pink Book

THE OBLIGATIONS OF THE CONTRACTOR (continued)

CLAUSE	OBLIGATIONS	TIME FRAME	SPECIFIC CONSEQUENCES OF NON-COMPLIANCE
17 Risk and Responsibility			
17.1 Indemnities	Indemnify and hold harmless the Employer's Personnel, and their respective agents against and from all claims, damages, losses and expenses in respect of bodily injury, sickness, disease, death, damage to or loss of property by reason of the Contractor's design, the execution and completion of the Works or negligence, wilful act or breach of the Contract by the Contractor, the Contractor's Personnel or agents.	None.	None.
17.2 Contractor's Care of the Works	a) Take full responsibility for the care of the Works and Goods. b) Take responsibility for the care of any work, which is outstanding on the date stated in a Taking-Over Certificate. c) Rectify loss or damage, if any loss or damage happens to the Works, Goods or Contractor's Documents.	a) From the Commencement Date until the Taking-Over Certificate is issued. b) Until the outstanding work has been completed. c) None.	None.
17.4 Consequences of Employer's Risks	a) Give notice in the case of an Employer's risk event, which results in loss or damage. b) Rectify the loss or damage as required by the Engineer. c) Give further notice if the Contractor suffers delay and/or incurs Cost as a result of rectifying loss or damage caused by Employers Risks.	a) Promptly. b) None. c) As soon as practicable and not later than 28 days after the Contractor became aware, or should have become aware of the event or circumstance (Sub-Clause 20.1).	a) None. b) None. c) Loss of entitlement to an extension to the Time for Completion and additional payment (Sub-Clause 20.1).

CLAUSE	OBLIGATIONS	TIME FRAME	SPECIFIC CONSEQUENCES OF NON-COMPLIANCE
17.5 Intellectual and Industrial Property Rights	e) Give notice of any claim under this clause. f) Indemnify and hold the Employer harmless against and from any other claim which arises out of, or in relation to the manufacture, use, sale or import of any goods, or any design for which the Contractor is responsible. g) If requested by the Employer, assist in contesting the claim. h) Not to make any admission which might be prejudicial to the Employer.	a) Within 28 days of receiving a claim. b) None. c) None. d) None.	a) Waiver of right to indemnity. b) None. c) None. d) None.
18 Insurance			
18.1 General Requirements for Insurances	a) Wherever the Contractor is the insuring Party, effect and maintain the insurances in terms consistent with any terms agreed by the Parties before the date of the Letter of Acceptance. b) Act under the policy on behalf of any additional joint insured parties. c) Submit evidence that the insurances have been effected and copies of the policies to the Employer. d) Submit evidence of payment of premiums. e) Inform the insurers of any relevant changes to the execution of the Works and ensure that insurance is maintained. f) Not to make any material alteration to the terms of any insurance without approval of the Employer.	a) Within the periods stated in the Contract Data. b) None. c) Within the periods stated in the Appendix to Tender. d) Upon payment of premium. e) As appropriate. f) None	Employer may effect the insurances and recover the cost from the Contractor.
18.4 Insurance for Contractor's Personnel	Effect and maintain insurance against injury, sickness, disease or death of any person employed by the Contractor, or any other of the Contractor's Personnel.	From the time that the personnel are assisting in the execution of the Works.	Employer may effect the insurances and recover the cost from the Contractor (Sub-Clause 18.1).

Pink Book

THE OBLIGATIONS OF THE CONTRACTOR (continued)

CLAUSE	OBLIGATIONS	TIME FRAME	SPECIFIC CONSEQUENCES OF NON-COMPLIANCE
19 Force Majeure			
19.2 Notice of Force Majeure	Give notice in the case that the Contractor is, or will be prevented from performing any of its obligations under the Contract by Force Majeure.	Within 14 days after the Contractor became aware, or should have become aware of, the relevant event or circumstance constituting Force Majeure.	Contractor shall not be excused performance of the obligations.
19.3 Duty to Minimise Delay	a) Use all reasonable endeavours to minimise any delay in the performance of the Contract as a result of Force Majeure. b) Give notice when the effects of the Force Majeure cease.	None.	None.
20 Claims, Disputes and Arbitration			
20.1 Contractor's Claims	a) Give notice if the Contractor considers himself to be entitled to any extension of the Time for Completion and/ or any additional payment. b) Submit any other notices which are required by the Contract and supporting particulars for the claim. c) Keep such contemporary records as may be necessary to substantiate any claim and permit the Engineer to inspect all the records. d) Send to the Engineer a fully detailed claim. e) Send further interim claims if the event or circumstance giving rise to the claim has a continuing effect. f) Send a final claim.	a) As soon as practicable and not later than 28 days after the Contractor became aware, or should have become aware of the event or circumstance. b) None. c) None. d) Within 42 days after the Contractor became aware, or should have become aware, of the event or circumstance giving rise to the claim. e) At monthly intervals. f) Within 28 days after the end of the effects resulting from the event or circumstance.	a,b,d) Loss of entitlement to an extension to the Time for Completion and additional payment. c,e,f) The Employer will take account of the extent to which the failure has prevented or prejudiced proper investigation of the claim.

CLAUSE	OBLIGATIONS	TIME FRAME	SPECIFIC CONSEQUENCES OF NON-COMPLIANCE
20.2 Appointment of the Dispute Board	a) Jointly appoint the Dispute Board (DB). b) Mutually agree the terms of remuneration for the DB. c) Not to consult the DB on any matter without the agreement of the other Party. d) Not to act alone in the termination of any member of the DB.	a) By the date stated in the Contract Data. b) None. c) None. d) None.	a) The appointing entity or official named in the Appendix to Tender shall appoint (Sub-Clause 20.3). b) None. c) None. d) None.
20.4 Obtaining Dispute Board's Decision	a) Make available to the DB, additional information, access to the Site and appropriate facilities as the DB may require. b) Give effect to a DB decision unless and until it is revised in an amicable settlement or an arbitral award. c) Continue to proceed with the Works in accordance with the Contract.	a) Promptly. b) Promptly. c) None.	a) None. b&c) The matter may be referred to arbitration (Sub-Clause 20.7).
20.5 Amicable Settlement	In the case of a Notice of Dissatisfaction being given with the DB's decision, attempt to settle the dispute amicably.	Within 56 days of the notice.	Arbitration may be commenced.

Pink Book

THE OBLIGATIONS OF THE CONTRACTOR (continued)

CLAUSE	OBLIGATIONS	TIME FRAME	SPECIFIC CONSEQUENCES OF NON-COMPLIANCE
GENERAL CONDITIONS OF DISPUTE BOARD AGREEMENT			
5 General Obligations of the Employer and the Contractor	a) Not to request advice from or consult with the Member regarding the Contract, otherwise than in the normal course of the DB's activities. b) In the case of the DB Member being required to make a site visit or attend a hearing, provide appropriate security for a sum equivalent to the reasonable expenses to be incurred by the Member (this may be undertaken by the Employer).	None.	None.
6 Payment	a) Pay the DB fees. b) Apply to the Employer for reimbursement of one-half of the DB invoices by way of the Statements.	a) Within 56 calendar days after receiving each invoice. b) None.	a) Employer entitled to pay the DB and to the reimbursement of fees, plus financing charges. The DB Member may suspend services or resign the appointment. b) The Employer is not obliged to reimburse the Contractor.

CLAUSE	OBLIGATIONS	TIME FRAME	SPECIFIC CONSEQUENCES OF NON-COMPLIANCE
Annex – Procedural Rules			
2.	Jointly agree the timing of and agenda for each Site visit by the DB.	None.	The DB shall decide the timing and agenda.
3.	a) Attend Site visits by the DB. b) Co-operate with the Employer in the co-ordination of Site visits by the DB. c) Provide the DB with conference facilities and secretarial and copying services.	None.	None.
4.	a) Furnish to each DB Member, one copy of all documents, which the DAB may request. b) Copy the Employer on all communications between the DB and the Contractor.	None.	None.

Pink Book

THE OBLIGATIONS OF THE ENGINEER

CLAUSE	OBLIGATIONS	TIME FRAME	SPECIFIC CONSEQUENCES OF NON-COMPLIANCE
GENERAL CONDITIONS			
1 General Provisions			
1.3 Communications	Not to unreasonably withhold approvals, certificates, consents and determinations.	None.	None.
1.5 Priority of Documents	In the case that an ambiguity or discrepancy is found in the Contract documents, issue any necessary clarification or instruction.	None.	None.
1.9 Delayed Drawings or Instructions	In the case of a notice and claim for delay or Cost being received, respond to the claim and agree or determine the matters.	Respond within 42 days after receiving a claim or any further particulars supporting a previous claim (Sub-Clause 20.1).	None.
2 The Employer			
2.1 Right of Access to the Site	In the case of a notice and claim for delay or Cost being received, respond to the claim and agree or determine the matters.	Respond within 42 days after receiving a claim or any further particulars supporting a previous claim (Sub-Clause 20.1).	None.
2.5 Employer's Claims	a) In the case that the Employer considers himself to be entitled to any payment, give notice and particulars to the Contractor (the Employer may also undertake this action). b) Agree or determine the matters.	a) As soon as practicable after the Employer became aware of the event or circumstances giving rise to the claim. b) Within 28 days from receipt of the claim (Sub-Clause 3.5).	None.

CLAUSE	OBLIGATIONS	TIME FRAME	SPECIFIC CONSEQUENCES OF NON-COMPLIANCE
3 The Engineer			
3.1 Engineer's Duties and Authority	a) Carry out the duties assigned in the Contract. b) Provide staff that are suitably qualified engineers and other professionals who are competent to carry out these duties. c) Obtain the approval of the Employer before exercising any authority specified in the Particular Conditions. d) Notify the Contractor of any act by the Engineer in response to a Contractor's request. e) Obtain approval of the Employer before taking the listed actions under Sub-Clauses 4.12, 13.1, 13.3 and 13.4.	a) None. b) None. c) None d) Within 28 days of receipt of notice (Sub-Clause 3.5). e) None.	a) None. b) None. c) The Employer shall be deemed to have given approval. d) None. e) None.
3.2 Delegation by the Engineer	a) In the case of delegation of the Engineer's authority, delegate such authority in writing. b) Not to delegate the authority to determine any matter in accordance with Sub-Clause 3.5 *[Determinations]*. c) In the case that the Contractor questions any determination or instruction of an assistant and refers the matter, confirm, reverse or vary the determination or instruction.	a) None. b) None. c) Promptly.	None.
3.3 Instructions of the Engineer	Wherever practical, give instructions in writing.	None.	None.
3.5 Determinations	a) Consult with each Party in an endeavour to reach an agreement. b) If agreement is not achieved, make a fair determination in accordance with the Contract, taking due regard of all relevant circumstances. c) Give notice to both Parties of each agreement or determination, with supporting particulars.	a) None. b) None c) Except where otherwise specified, within 28 days from the receipt of the corresponding claim or request.	None.

Pink Book

THE OBLIGATIONS OF THE ENGINEER (continued)

CLAUSE	OBLIGATIONS	TIME FRAME	SPECIFIC CONSEQUENCES OF NON-COMPLIANCE
4 The Contractor			
4.3 Contractor's Representative	Respond to the Contractor's request for consent to the appointment of the Contractor's Representative.	None.	None.
4.4 Subcontractors	Respond to the Contractor's request for consent to the appointment of Subcontractors.	None.	None.
4.7 Setting Out	In the case of a notice and claim for Cost or delay being received, respond to the claim and agree or determine the matters.	Respond within 42 days after receiving a claim or any further particulars supporting a previous claim (Sub-Clause 20.1).	None.
4.12 Unforeseeable Physical Conditions	a) In the case of a notice of unforeseen physical conditions, inspect the physical conditions. b) In the case of a notice and claim for cost or delay being received, respond to the claim and agree or determine the matters.	a) None. b) Respond within 42 days after receiving a claim or any further particulars supporting a previous claim (Sub-Clause 20.1).	None.
4.19 Electricity, Water and Gas	Agree or determine the quantities and amounts due to the Employer for the Contractor's consumption of electricity, water and gas.	None.	None.
4.20 Employer's Equipment and Free-Issue Material	Agree or determine the quantities and the amounts due to the Employer for the Contractor's use of Employer's Equipment.	None.	None.
4.23 Contractor's Operations on Site	Cooperate with the Contractor to agree additional working areas outside the Site.	None.	None.

CLAUSE	OBLIGATIONS	TIME FRAME	SPECIFIC CONSEQUENCES OF NON-COMPLIANCE
4.24 Fossils	a) Give instructions for dealing with fossils, coins, articles of value or antiquity, and structures and other remains or items of geological or archaeological interest found on the Site. b) In the case of a notice and claim for delay or Cost being received, respond to the claim and agree or determine the matters.	a) None. b) Respond within 42 days after receiving a claim or any further particulars supporting a previous claim (Sub-Clause 20.1).	None.
5 Nominated Subcontractors			
5.3 Payments to Nominated Subcontractors	Certify the amounts due to nominated Subcontractors.	None.	None.
6 Staff and Labour			
6.10 Records of Contractor's Personnel and Equipment	Cooperate with the Contractor to approve a form to record the number of each class of Contractor's Personnel and of each type of Contractor's Equipment on the Site.	None.	None.
7 Plant, Materials and Workmanship			
7.3 Inspection	Examine, inspect, measure and test the materials and workmanship.	Without unreasonable delay (or promptly give notice that inspection is not required).	None.

Pink Book

THE OBLIGATIONS OF THE ENGINEER (continued)

CLAUSE	OBLIGATIONS	TIME FRAME	SPECIFIC CONSEQUENCES OF NON-COMPLIANCE
7.4 Testing	a) Agree with the Contractor, the time and place for specified testing. b) Give the Contractor notice of intention to attend the tests. c) In the case of a notice and claim for delay or Cost being received, respond to the claim and agree or determine the matters. d) Endorse the Contractor's test certificate, or issue a certificate confirming tests have been passed.	a) None. b) Not less than 24 hours. c) Respond within 42 days after receiving a claim or any further particulars supporting a previous claim (Sub-Clause 20.1). d) Promptly.	a) None. b) If the Engineer does not attend, the Contractor may proceed and the tests shall be deemed to have been made in the Engineer's presence. c) None. d) None.
7.5 Rejection	In the case of Plant, Materials or workmanship being found to be defective or not in accordance with the Contract, give notice of rejection.	None.	None.
8 Commencement, Delays and Suspension			
8.1 Commencement of Works	Give the Contractor instructions to commence work on fulfilment of the listed conditions.	Within 180 days of receipt of the Letter of Acceptance by the Contractor.	The Contractor is entitled to terminate the Contract.
8.3 Programme	In the case that a programme does not comply with the Contract, give notice to the Contractor.	Within 21 days after receiving the programme.	Contractor shall proceed in accordance with the programme.
8.4 Extension of Time for Completion	In the case of a notice and claim for delay being received, respond to the claim and agree or determine the matters.	Respond within 42 days of receiving a claim or any further particulars supporting a previous claim (Sub-Clause 20.1).	None.

CLAUSE	OBLIGATIONS	TIME FRAME	SPECIFIC CONSEQUENCES OF NON-COMPLIANCE
8.9 Consequences of Suspension	In the case of a notice and claim for delay or Cost being received, respond to the claim and agree or determine the matters.	Respond within 42 days after receiving a claim or any further particulars supporting a previous claim (Sub-Clause 20.1).	None.
8.11 Prolonged Suspension	In the case that the Contractor requests permission to proceed after 84 days of suspension, respond to the Contractor's request.	Within 28 days of the request (Sub-Clause 3.5).	Contractor may treat the suspension as an omission or give notice of termination.
8.12 Resumption of Work	Jointly examine the Works and the Plant and Materials affected by the suspension.	After permission or instruction to proceed is given.	None.
9 Tests on Completion			
9.1 Contractor's Obligations	Make allowances for the effect of any use of the Works by the Employer on the performance or other characteristics of the Works.	None.	None.
10 Employer's Taking Over			
10.1 Taking Over of the Works and Sections	Issue the Taking-Over Certificate to the Contractor, or reject the Contractor's application giving reasons and specifying the work required to be done.	Within 28 days after receiving the Contractor's application.	The Taking-Over Certificate shall be deemed to have been issued.

Pink Book

THE OBLIGATIONS OF THE ENGINEER (continued)

CLAUSE	OBLIGATIONS	TIME FRAME	SPECIFIC CONSEQUENCES OF NON-COMPLIANCE
10.2 Taking Over of Parts of the Works	a) In the case that the Employer uses part of the Works and if requested by the Contractor, issue a Taking-Over Certificate for this part. b) In the case of a notice and claim for incurred Cost being received, respond to the claim and agree or determine the matters. c) Determine any reduction in delay damages as a result of a Taking-Over Certificate being issued for a part of the Works.	a) None. b) Respond within 42 days after receiving a claim or any further particulars supporting a previous claim (Sub-Clause 20.1). c) None.	None.
10.3 Interference with Tests on Completion	a) In the case of the Contractor being prevented from carrying out Tests on Completion by the Employer, issue a Taking-Over Certificate accordingly. b) In the case of a notice and claim for delay or cost being received, respond to the claim and agree or determine the matters.	a) None. b) Respond within 42 days after receiving a claim or any further particulars supporting a previous claim (Sub-Clause 20.1).	None.
11 Defects Liability			
11.4 Failure to Remedy Defects	In the case that the Contractor fails to remedy any defect and if required by the Employer, agree or determine a reasonable reduction in theContract Price.	None.	None.
11.8 Contractor to Search	In the case that the Contractor has searched for a defect that is found not to be the responsibility of the Contractor, agree or determine the cost of the search.	Within 28 days of a request to do so (Sub-Clause 3.5).	None.

CLAUSE	OBLIGATIONS	TIME FRAME	SPECIFIC CONSEQUENCES OF NON-COMPLIANCE
11.9 Performance Certificate	Issue the Performance Certificate.	Within 28 days after the latest of the expiry dates of the Defects Notification Periods or as soon thereafter as the Contractor has completed his obligations.	None.
12 Measurement and Evaluation			
12.1 Works to be Measured	a) Give notice when the Engineer requires any part of the Works to be measured. b) Prepare records of measurement. c) In the case that the Contractor gives notice or disagrees with the records, review the records and either confirm or vary them.	a) None. b) None. c) Within 28 days of the notice (Sub-Clause 3.5).	None.
12.3 Evaluation	a) Agree or determine the Contract Price by evaluating each item of work, applying the measurement and the appropriate rate or price for the item. b) In the case that new rates and prices are required and not agreed, determine a provisional rate or price for the purposes of Interim Payment Certificates.	a) None. b) As soon as the concerned work commences.	None.
12.4 Omissions	In the case of receiving a notice of cost as a result of omissions, agree or determine the cost.	Within 28 days of the notice (Sub-Clause 3.5).	None.
13 Variations and Adjustments			
13.1 Right to Vary	In the case that the Contractor gives notice that the Contractor cannot readily obtain the Goods required for a Variation, cancel, confirm or vary the instruction.	None.	None.

Pink Book

THE OBLIGATIONS OF THE ENGINEER (continued)

CLAUSE	OBLIGATIONS	TIME FRAME	SPECIFIC CONSEQUENCES OF NON-COMPLIANCE
13.2 Value Engineering	In the case that a proposal results in a reduction in the contract value, agree or determine a fee.	Within 28-days (Sub-Clause 3.5).	None.
13.3 Variation Procedure	a) Respond to the Contractor's Variation or Value Engineering proposals with approval, disapproval or comments. b) Issue instructions to execute Variations.	a) As soon as practicable after receiving the proposal. b) None.	a) None. b) The Contractor shall not make any alteration and/or modification of the Permanent Works (Sub-Clause 13.1).
13.5 Provisional Sums	Give instructions for the use of Provisional Sums.	None.	None.
13.6 Daywork	Sign the Contractor's Daywork statements.	If correct, or when agreed.	None.
13.7 Adjustments for Changes in Legislation	In the case of a notice and claim for delay being received, respond to the claim and agree or determine the matters.	Respond within 42 days after receiving a claim or any further particulars supporting a previous claim (Sub-Clause 20.1).	None.
13.8 Adjustments for Changes in Cost	a) In the case that the cost indices or reference prices stated in the table of adjustment data is in doubt, make a determination. b) In the case that each current cost index is not available, determine a provisional index for the issue of Interim Payment Certificates.	a) None. b) Such that the index may be used for calculations for inclusion in the Payment Certificates.	None.

CLAUSE	OBLIGATIONS	TIME FRAME	SPECIFIC CONSEQUENCES OF NON-COMPLIANCE
14 Contract Price and Payment			
14.2 Advance Payment	Issue an Interim Payment Certificate for the advance payment or its first installment.	After receiving a Statement under Sub-Clause 14.3 and after the Employer receives the Performance Security and an advance payment guarantee.	None.
14.3 Application for Interim Payment Certificates	Cooperate with the Contractor to agree and approve the form for the Statements.	None.	None.
14.5 Plant and Materials intended for the Works	Determine and certify an amount for Plant and Materials which have been sent to the Site for incorporation in the Permanent Works.	For inclusion in each Interim Payment Certificate.	None.
14.6 Issue of Interim Payment Certificates	a) Issue to the Employer and the Contractor an Interim Payment Certificate. b) In the case that the certified amount would be less than the minimum amount of Interim Payment Certificates stated in the Contract Data, give notice to the Contractor.	a) Within 28 days after receiving a Statement from the Contractor. b) None.	If late certification results in the Employer not making payment within the stated period, the Contractor is entitled to receive financing charges (Sub-Clause 14.7).
14.9 Payment of Retention Money	a) Certify the first half of the Retention Money. b) Certify the outstanding balance of the Retention Money.	a) When the Taking-Over Certificate has been issued for the Works. b) Promptly after the latest of the expiry dates of the Defects Notification Periods.	If late certification results in the Employer not making payment within the stated period, the Contractor is entitled to receive financing charges (Sub-Clause 14.7).

THE OBLIGATIONS OF THE ENGINEER

Pink Book

THE OBLIGATIONS OF THE ENGINEER (continued)

CLAUSE	OBLIGATIONS	TIME FRAME	SPECIFIC CONSEQUENCES OF NON-COMPLIANCE
14.10 Statement at Completion	Issue to the Employer an Interim Payment Certificate.	Within 28 days after receiving a Statement at Completion.	If late certification results in the Employer not making payment within the stated period, the Contractor is entitled to receive financing charges (Sub-Clause 14.7).
14.11 Application for Final Payment Certificate	a) Cooperate with the Contractor to agree and approve the form for the draft final statement. b) In the case that a dispute exists, deliver to the Employer (with a copy to the Contractor) an Interim Payment Certificate for the agreed parts of the draft final statement.	None.	None.
14.13 Issue of Final Payment Certificate	a) Issue, to the Employer the Final Payment Certificate. b) In the case that the Contractor has not applied for a Final Payment Certificate, request the Contractor to do so. c) In the case that the Contractor fails to submit an application within a period of 28 days, issue the Final Payment Certificate for such amount as the Engineer fairly determines to be due.	a) Within 28 days after receiving the Final Statement and written discharge. b) None. c) None.	If late certification results in the Employer not making payment within the stated period, the Contractor is entitled to receive financing charges (Sub-Clause 14.7).

CLAUSE	OBLIGATIONS	TIME FRAME	SPECIFIC CONSEQUENCES OF NON-COMPLIANCE
15 Termination by Employer			
15.3 Valuation at Date of Termination	Agree or determine the value of the Works, Goods, Contractor's Documents and any other sums due to the Contractor for work executed in accordance with the Contract.	As soon as practicable after a notice of termination.	None.
16 Suspension and Termination by Contractor			
16.1 Contractor's Entitlement to Suspend Work	In the case of a notice and claim for delay or Cost being received, respond to the claim and agree or determine the matters.	Respond within 42 days after receiving a claim or any further particulars supporting a previous claim (Sub-Clause 20.1).	None.
17 Risk and Responsibility			
17.4 Consequences of Employer's Risks	In the case of a notice and claim for delay or Cost being received, respond to the claim and agree or determine the matters.	Respond within 42 days after receiving a claim or any further particulars supporting a previous claim (Sub-Clause 20.1).	None.
19 Force Majeure			
19.4 Consequences of Force Majeure	In the case of a notice and claim for delay or Cost being received, respond to the claim and agree or determine the matters.	Respond within 42 days after receiving a claim or any further particulars supporting a previous claim (Sub-Clause 20.1).	None.
19.6 Optional Termination, Payment and Release	In the case of termination, determine the value of the work done and issue a Payment Certificate.	Upon termination.	None.

THE OBLIGATIONS OF THE ENGINEER

Pink Book

THE OBLIGATIONS OF THE ENGINEER (continued)

CLAUSE	OBLIGATIONS	TIME FRAME	SPECIFIC CONSEQUENCES OF NON-COMPLIANCE
20 Claims, Disputes and Arbitration			
20.1 Contractor's Claims	a) In the case of a claim being received, respond with approval, or with disapproval and detailed comments. b) Agree or determine the extension of the Time for Completion and/or the additional payment.	Within 42 days after receiving the claim or any further particulars supporting a previous claim (Sub-Clause 20.1).	The Parties may consider that the claim is rejected and refer the matter to the Dispute Board.
GENERAL CONDITIONS OF DISPUTE BOARD AGREEMENT			
Annex – Procedural Rules			
3.	Attend Site visits by the DB.	None.	None.

THE OBLIGATIONS OF THE DISPUTE BOARD

CLAUSE	OBLIGATIONS	TIME FRAME	SPECIFIC CONSEQUENCES OF NON-COMPLIANCE
GENERAL CONDITIONS			
20 Claims, Disputes and Arbitration			
20.4 Obtaining Dispute Board's Decision	Give a decision on any dispute referred to the DB.	Within 84 days after receiving such reference.	Either Party may commence arbitration.
GENERAL CONDITIONS OF DISPUTE BOARD AGREEMENT			
3 Warranties	a) Be impartial and independent of the Employer, the Contractor and the Engineer. b) Disclose to the Parties and to the Other Members, any fact or circumstance which might appear inconsistent with his/her warranty and agreement of impartiality and independence.	a) None. b) Promptly.	None.

Pink Book

THE PINK BOOK: CONDITIONS OF CONTRACT FOR CONSTRUCTION, MDB HARMONISED EDITION

THE OBLIGATIONS OF THE DISPUTE BOARD (continued)

CLAUSE	OBLIGATIONS	TIME FRAME	SPECIFIC CONSEQUENCES OF NON-COMPLIANCE
4 General Obligations of the Member	a) Have no interest, financial or otherwise in the Parties or the Engineer, nor any financial interest in the Contract. b) Not previously to have been employed as a consultant or otherwise by the Parties or the Engineer, except as disclosed in writing. c) Disclose in writing to the Parties and the Other Members, any professional or personal relationships with any director, officer or employee of the Parties or the Engineer and any previous involvement in the overall project of which the Contract forms part. d) Not, for the duration of the Dispute Board Agreement, be employed as a consultant or otherwise by the Parties or the Engineer, except as may be agreed in writing. e) Comply with the procedural rules and with Sub-Clause 20.4 of the Conditions of Contract. f) Not give advice to the Parties, the Employer's Personnel or the Contractor's Personnel concerning the conduct of the Contract, other than in accordance with the procedural rules. g) Not enter into discussions or make any agreement with the Employer, the Contractor or the Engineer regarding employment by any of them after ceasing to act under the Dispute Board Agreement. h) Ensure his/her availability for all site visits and hearings as are necessary. i) Become conversant with the Contract and with the progress of the Works by studying all documents received. j) Treat the details of the Contract and all the DB's activities and hearings as private and confidential. k) Be available to give advice and opinions on any matter relevant to the Contract when requested by both of the Parties.	None.	None.

CLAUSE	OBLIGATIONS	TIME FRAME	SPECIFIC CONSEQUENCES OF NON-COMPLIANCE
5 General Obligations of the Employer and the Contractor	a) Not be appointed as an arbitrator in any arbitration under the Contract.		None.
6 Payment	a) Submit invoices for payment of the monthly retainer and air fares. b) Submit invoices for other expenses and for daily fees.	a) Quarterly in advance. b) Following the conclusion of a site visit or hearing.	None.
Annex – Procedural Rules			
1.	Visit the Site.	a) At intervals of not more than 140 days. b) At times of critical construction events. c) At the request of either the Employer or the Contractor.	None.
2.	a) Agree the timing of and agenda for each Site visit with the Parties. b) In the absence of agreement by the Parties, decide the timing of and agenda for each Site visit.	None.	None.
3.	Prepare a report on the DB's activities during the visit and send copies to the Employer and the Contractor.	At the conclusion of each site Site visit and before leaving the site.	None.
4.	Copy all communications to the Parties.	None.	None.
5(a).	a) Act fairly and impartially as between the Parties. b) Give each of the Parties a reasonable opportunity of putting his case and responding to the other's case.	None.	None.
5(b).	Adopt procedures suitable to the dispute, avoiding unnecessary delay or expense.	None.	None.

Pink Book

THE OBLIGATIONS OF THE DISPUTE BOARD (*continued*)

CLAUSE	OBLIGATIONS	TIME FRAME	SPECIFIC CONSEQUENCES OF NON-COMPLIANCE
6.	In the case of a hearing on the dispute, decide on the date and place for the hearing.	None.	None.

9.	a) Not express any opinions during any hearing concerning the merits of any arguments advanced by the Parties.
	b) Make and give a decision in accordance with Sub-Clause 20.4, or as otherwise agreed by the Employer and the Contractor in writing.
	c) If the DB comprises three persons:
	  I. Convene in private after a hearing.
	  II. Endeavour to reach a unanimous decision.
		None.	None.

Chapter 3

The Red Book Subcontract

Conditions of Subcontract for Construction for Building and Engineering Works Designed by the Employer, First Edition 2011

The FIDIC Contracts: Obligations of the Parties, First Edition. Andy Hewitt.
© 2014 John Wiley & Sons, Ltd. Published 2014 by John Wiley & Sons, Ltd.

THE OBLIGATIONS OF THE CONTRACTOR

CLAUSE	OBLIGATIONS	TIME FRAME	SPECIFIC CONSEQUENCES OF NON-COMPLIANCE
	**GENERAL CONDITIONS**		
1 Definitions and Interpretation			
1.5 Priority of Subcontract Documents	a) In the case that an ambiguity or discrepancy is found in the Subcontract documents, issue any necessary clarification or Contractor's Instruction. b) In the case that the Contractor becomes aware of an error or defect in a document prepared for the Subcontract Works, give notice to the Subcontractor.	a) None. b) Promptly.	None.
1.6 Notices, Consents, Approvals, Certificates, Confirmations, Decisions and Determinations	Not to unreasonably withhold approvals, certificates, consents and determinations.	None.	None.
1.7 Joint and Several Liability under the Subcontract	a) In the case of a joint venture, consortium or other unincorporated grouping of two or more persons, notify the Subcontractor of the leader. b) Not to alter the composition or legal status of the joint venture without the prior consent of the Subcontractor.	None.	None.
1.9 Subcontract Agreement	Enter into and execute a Subcontract Agreement.	Within 28 days after the Subcontractor receives the Contractor's Letter of Acceptance, unless agreed otherwise.	None.

Red Book Subcontract

THE OBLIGATIONS OF THE CONTRACTOR (continued)

CLAUSE	OBLIGATIONS	TIME FRAME	SPECIFIC CONSEQUENCES OF NON-COMPLIANCE
2 The Main Contract			
2.1 Subcontractor's Knowledge of Main Contract	a) Make all the documents of the Main Contract available to the Subcontractor, except for prices and confidential parts. b) Provide a copy of the Appendix to Tender and the particular Conditions of the Main Contract. c) If requested, provide a true copy of the Main Contract, except for prices and confidential parts. d) If notice of an ambiguity or discrepancy is given, issue any necessary clarification or Contractor's Instruction.	None.	None.
2.4 Rights, Entitlements and Remedies under the Main Contract	Take all reasonable steps to secure from the Employer such rights, entitlements and remedies as the Contractor has under the Main Contract, with respect to the Subcontract Works.	None.	None.
3 The Contractor			
3.2 Access to the Site	Give the Subcontractor right of access to and possession of so much of the Site for the Subcontractor to proceed with execution of the Subcontract Works.	Within the times required in accordance with the Subcontract Programme.	None.

CLAUSE	OBLIGATIONS	TIME FRAME	SPECIFIC CONSEQUENCES OF NON-COMPLIANCE
3.3 Contractor's Claims in connection with the Subcontract	a) In the case that the Contractor considers himself to be entitled to any payment under any Clause of the Conditions or otherwise in connection with the Subcontract, give notice specifying the basis of the claim to the Subcontractor. b) Send detailed particulars of the claim including substantiation of the amount. c) Consult with the Subcontractor in an endeavour to reach agreement on the amount. d) In the case that agreement is not reached, make a fair decision on the amount. e) Give notice with reasons and supporting particulars of the decision. f) Pay financing charges in respect of any deduction from sums otherwise due to the Subcontractor of an amount to which the Contractor was not entitled.	a) As soon as practicable and not later than 28 days after becoming aware of the event or circumstances giving rise to the claim. b) As soon as practicable and not less than 28 days after giving notice. c) None. d) None. e) None. f) None.	None.
3.4 Employer's Claims in connection with the Main Contract	In the case that the Contractor receives any notice and particulars of an Employer's claim which concerns the Subcontractor, send a copy to the Subcontractor.	Immediately.	None.
3.5 Co-ordination of Main Works	a) Be responsible for overall co-ordination and project management of the Main Works. b) Be responsible for co-ordination of the Subcontract Works with the works of the Contractor and other subcontractors.	None.	None.

**Red Book
Subcontract**

THE OBLIGATIONS OF THE CONTRACTOR (continued)

CLAUSE	OBLIGATIONS	TIME FRAME	SPECIFIC CONSEQUENCES OF NON-COMPLIANCE
4 The Subcontractor			
4.2 Subcontract Performance Security	Not to unreasonably withhold or delay approval of the form of the Subcontract Performance Security.	None.	None.
5 Assignment of the Subcontract and Subcontracting			
5.1 Assignment of Subcontract	Not to assign the whole or any part of the Subcontract or any benefit or interest in or under the Subcontract, without prior consent of the Subcontractor, except for instances listed in this Sub-Clause.	None.	None.
6 Co-operation, Staff and Labour			
6.1 Co-operation under the Subcontract	Ensure that the Contractor's Personnel and any other Subcontractors co-operate with and allow appropriate opportunities for carrying out work to the Subcontractor.	None.	In the case of delay and/or the incurrence of Cost, the Subcontractor is entitled to an extension of time and/or additional payment.
6.2 Persons in the Service of Others	Not to recruit, or attempt to recruit, staff and labour from amongst the Subcontractor's Personnel.	None.	None.

CLAUSE	OBLIGATIONS	TIME FRAME	SPECIFIC CONSEQUENCES OF NON-COMPLIANCE
6.3 Contractor's Subcontract Representative	a) Appoint the Contractor's Subcontract Representative and give him all authority necessary to act on the Contractor's behalf under the Subcontract. b) Notify the Subcontractor of the name and particulars of the Contractor's Subcontract Representative. c) In the case that the Contractor's Subcontract Representative is to be temporarily absent, notify the Subcontractor of a replacement person. d) Ensure that the Contractor's Subcontract Representative is fluent in the language for communications.	a) None. b) Prior to the Subcontract Commencement Date. c) As appropriate. d) None.	None.
7 Equipment, Temporary Works, Other Facilities, Plant and Materials			
7.1 Subcontractor's Use of Equipment, Temporary Works, and/or Other Facilities	a) Make the Employer's Equipment, the Contractor's Equipment, the Temporary Works and/or other facilities specified available to the Subcontractor. b) In the case that the Subcontractor gives notice of any shortage, defect or default, rectify the matter.	a) Within the times required to enable the Subcontractor to proceed in accordance with the Subcontract Programme. b) Immediately.	None.
7.2 Free-Issue Materials	a) Supply, free of charge, the free-issue materials specified. b) In the case that the Subcontractor gives notice of any shortage, defect or default, rectify the matter.	a) Within the times required to enable the Subcontractor to proceed in accordance with the Subcontract Programme. b) Immediately	None.
8 Commencement and Completion			
8.1 Commencement of Subcontract Works	Notify the Subcontractor of the Subcontract Commencement Date.	Not less than 14 days before the Subcontract Commencement Date.	None.

THE OBLIGATIONS OF THE CONTRACTOR

Red Book
Subcontract

THE OBLIGATIONS OF THE CONTRACTOR (continued)

CLAUSE	OBLIGATIONS	TIME FRAME	SPECIFIC CONSEQUENCES OF NON-COMPLIANCE
8.3 Subcontract Programme	Give the Subcontractor all reasonable co-operation and assistance in order that he may progress the Subcontract Works as required by the Subcontract Programme.	None.	None.
8.5 Subcontract Progress Reports	If monthly progress reports are required by the Contractor, notify the Subcontractor of the due date for the submission of Contractor's monthly progress reports	As appropriate.	None.
8.6 Suspension of the Subcontract Works by the Contractor	a) In the case of issuing a Contractor's instruction for the suspension of the Subcontract Works, state the reason or reasons for the suspension. b) In the case of suspension of all or part of the Subcontract Works, comply with any obligations detailed under the Main Contract Clauses 8.9, 8.10, 8.11 and 8.12.	a) None. b) As appropriate.	a) None. b) As appropriate.
10 Completion and Taking Over the Subcontract Works			
10.1 Completion of Subcontract Works	When the Subcontractor notifies that, in the Subcontractor's opinion, that the Subcontract Works will be complete: a) Notify the Subcontractor that completion of the Subcontract Works has been achieved, stating the date of completion; or b) Notify the Subcontractor that completion of the Subcontract Works has been not been achieved, giving reasons and specifying the work required to achieve completion.	Within 21 days of the Subcontractor's notice that the Subcontract Works will be complete.	None.

CLAUSE	OBLIGATIONS	TIME FRAME	SPECIFIC CONSEQUENCES OF NON-COMPLIANCE
11 Defects Liability			
11.1 Subcontractor's Obligations after Taking-Over	a) In the case that the Subcontractor fails to remedy notified defects or damage, and the remedial work was to be executed by the Employer, and the Engineer determines a reasonable reduction in the Main Contract Price, notify the Subcontractor. b) Comply with any obligations detailed under Main Contract Clauses 11.1, 11.4, 11.5, 11.6, 11.7 and 11.8.	a) Promptly. b) As appropriate.	a) None. b) As appropriate.
11.3 Performance Certificate	Forward a copy of the Performance Certificate to the Subcontractor.	Immediately on receipt of the Performance Certificate.	None.
12 Measurement and Evaluation			
12.1 Measurement of Subcontract Works	a) Permit the Subcontractor to attend to assist the Engineer and the Contractor in measurement in relation to the Subcontract Works. b) In the case that the Subcontract Works are to be measured by records, permit the Subcontractor to attend with the Contractor to examine and agree the records with the Engineer. c) In the case that the Subcontractor submits a notice of disagreement, notify the Engineer. d) Notify the Subcontractor of any determination made by the Engineer in respect of disagreed records. e) In the case that the Contractor does not give notice, or the timing of the notice is unreasonable, consult with the Subcontractor in an endeavour to reach agreement on the measurement of the Subcontract Works. f) If agreement is not reached, make a fair decision and give notice of the decision.	a) None. b) None. c) None. d) Immediately. e) None. f) None.	None.

Red Book
Subcontract

THE OBLIGATIONS OF THE CONTRACTOR (continued)

CLAUSE	OBLIGATIONS	TIME FRAME	SPECIFIC CONSEQUENCES OF NON-COMPLIANCE
12.3 Evaluation under the Subcontract	a) Consult with the Subcontractor in an endeavour to reach agreement on the Subcontract Price in accordance with the Main Contract Clauses 12.1 and 12.3. b) If agreement is not reached, make a fair evaluation and notify the Subcontractor with supporting particulars. c) Give effect to each agreement reached.	a) None. b) Promptly. c) Until revised under Clause 20 (Claims and Disputes).	None.
13 Subcontract Variations and Adjustments			
13.3 Request for Proposal for Subcontract Variation	Respond to the Subcontractor's proposal with approval, disapproval or comments.	As soon as practicable after receiving the proposal.	None.
14 Subcontract Price and Payment			
14.2 Subcontract Advance Payment	Make the advance payment in the instalments and in the applicable currencies and proportions stated in the Appendix to the Subcontractor's Offer.	None.	1. Contractor shall pay financing charges to the Subcontractor (Sub-Clause 14.9). 2. Subcontractor may suspend or reduce the rate of work (Sub-Clause 16.1).

CLAUSE	OBLIGATIONS	TIME FRAME	SPECIFIC CONSEQUENCES OF NON-COMPLIANCE
14.5 Contractor's Application for Interim Payment Certificate	a) Make appropriate provision for the amounts set out in the Subcontractor's monthly statement in the Contractor's Statement. b) If so requested, advise the Subcontractor of the dates that the Contractor's next Statement was submitted to the Engineer and the Contractor received an Interim Payment Certificate.	a) In the Contractor's next Statement. b) None.	The Interim Payment Certificate shall be deemed to have been issued 35 days after the due date for submission of the Contractor's Statement.
14.6 Interim Subcontract Payments	a) Pay the Subcontractor the amounts included in the Subcontractor's monthly statement and any other sums to which, in the Contractor's opinion, the Subcontractor is entitled. b) In the case that the Contractor is entitled to withhold or defer payment of any sums in a Subcontractor's monthly statement, give notice to the Subcontractor with reasons, full particulars and substantiating documents. c) Pay the Subcontractor any amount in a Subcontractor's monthly statement which has previously been withheld or deferred.	a) Within 70 days of receipt by the Contractor of the Subcontractor's monthly statement or Statement at Completion. b) Within 70 days of receipt by the Contractor of the Subcontractor's monthly statement. c) Within 7 days of receipt by the Contractor of any payment which includes a sum in respect of this amount.	1. Contractor shall pay financing charges to the Subcontractor (Sub-Clause 14.9). 2. Subcontractor may suspend or reduce the rate of work (Sub-Clause 16.1).
14.7 Payment of Retention Money under the Subcontract	a) Pay the Subcontractor the retention money under the Subcontract in the same proportions that apply under the Main Contract. b) Pay the Subcontractor the outstanding balance of the retention money under the subcontract.	a) No later than 14 days after the Contractor has received payment. b) No later than 7 days after the expiry of the Subcontract Defects Notification Period.	1. Contractor shall pay financing charges to the Subcontractor (Sub-Clause 14.9). 2. Subcontractor may suspend or reduce the rate of work (Sub-Clause 16.1).

THE OBLIGATIONS OF THE CONTRACTOR

Red Book Subcontract

THE OBLIGATIONS OF THE CONTRACTOR (continued)

CLAUSE	OBLIGATIONS	TIME FRAME	SPECIFIC CONSEQUENCES OF NON-COMPLIANCE
14.8 Final Subcontract Payment	Pay the Subcontractor the balance of the Subcontract Price finally due.	Within 56 days after the expiry of the Subcontract Defects Notification Period.	Contractor shall pay financing charges to the Subcontractor (Sub-Clause 14.9).
14.10 Cessation of the Contractor's Liability	Notify the Subcontractor of the date stated in the Performance Certificate issued under the Main Contract.	Promptly.	None.
14.11 Subcontract Currencies of Payment	Pay the Subcontractor in the currency or currencies named in the Appendix to the Subcontractor's Offer.	None.	None.
15 Termination of the Main Contract and Termination of the Subcontract by the Contractor			
15.1 Termination of Main Contract	a) Notify the Subcontractor of the date the Employer returns the Performance Security under the Main Contract. b) Return the Subcontract Performance Security to the Subcontractor.	a) Promptly. b) Within 7 days after the Employer returns the Performance Security under the Main Contract, or within 28 days after a notice of termination has taken effect, whichever is earlier.	None.
15.2 Valuation at Date of Subcontract Termination	a) Evaluate the Subcontract Works, Subcontract Goods, Subcontractor's Documents and any other sums due to the Subcontractor for work executed in accordance with the Subcontract. b) Give notice, with supporting particulars of this evaluation.	As soon as practicable after a notice of termination.	None.

CLAUSE	OBLIGATIONS	TIME FRAME	SPECIFIC CONSEQUENCES OF NON-COMPLIANCE
15.6 Termination of Subcontract by the Contractor	In the case that the Contractor wishes to terminate the Subcontract, give notice to the Subcontractor.	14 days before the date of termination.	None.
16 Suspension and Termination by the Subcontractor			
16.3 Payment on Termination by Subcontractor	a) Return the Subcontract Performance Bond. b) Pay the Subcontractor in accordance with Sub-Clause 15.3.	Promptly after a notice of termination has been issued by the Subcontractor.	None.
17 Risk and Indemnities			
17.2 Contractor's Indemnities	Indemnify and hold harmless the Subcontractor, the Contractor's personnel and their respective agents against and from all claims, damages, losses and expenses in respect of: a) Matters described in Main Contract Clause 18.3, sub-paragraph (d)(i) and (iii). b) Fault, error, defect or omission in any element of the Contractor's design. c) Bodily injury, disease or death which is attributable to any negligence, wilful act or breach of the Contract by the Employer, the Employer's Personnel, the Contractor and the Contractor's Personnel and any of their agents.	None.	None.
18 Subcontract Insurances			
18.2 Insurance arranged by the Contractor and/or the Employer	Effect and maintain the insurances for which the Contractor is responsible in accordance with the details set out in Annexe E.	None.	None.

THE OBLIGATIONS OF THE CONTRACTOR

Red Book
Subcontract

THE OBLIGATIONS OF THE CONTRACTOR (continued)

CLAUSE	OBLIGATIONS	TIME FRAME	SPECIFIC CONSEQUENCES OF NON-COMPLIANCE
18.3 Evidence of Insurance and Failure to Insure	If so requested by the Subcontractor, provide evidence of the insurance and payment of the current premium.	Promptly.	Subcontractor shall pay the premium and be entitled to reimbursement of the cost, plus expenses incurred.
20 Notices, Subcontractor's Claims and Disputes			
20.1 Notices	Give notice to the Subcontractor of any delay event which has occurred or specific probable future event(s) or circumstance(s) which may adversely affect the Subcontractor's activities or delay the execution of the Subcontract Works and/or the Main Works.	Immediately.	None.
20.2 Subcontractor's Claims	a) Consult with the Subcontractor in an endeavour to reach agreement on the extension of the Subcontract Time for Completion and/or additional payment. b) If agreement is not reached, make a fair decision, notify the Subcontractor with reasons, make the additional payment and grant the extension of the Subcontract Time for Completion. c) Give effect to each agreement reached or decision made unless and until revised.	a) None. b) Within 49 days of receiving the Subcontractor's fully detailed claim or any further particulars requested by the Contractor. c) As appropriate.	None.

CLAUSE	OBLIGATIONS	TIME FRAME	SPECIFIC CONSEQUENCES OF NON-COMPLIANCE
20.4 Subcontract Disputes	a) In the case that a dispute is involved in a dispute under the Main Contract and the Contractor has not previously referred the subject to the Main Contract DAB, refer the Subcontract dispute to the Main Contract DAB. b) If there is no DAB in place under the Main Contract, notify the Subcontractor.	a) Within 28 days. b) Immediately.	a) Subcontractor entitled to refer the dispute to the subcontract DAB. b) None.
20.5 Appointment of the Subcontract DAB	a) Jointly, with the Subcontractor, appoint the DAB. b) Not to act alone in the termination of any member of the Dispute Adjudication Board.	a) Within 42 days after the date of a Notice of Dispute. b) None.	a) The President of FIDIC or his appointee may appoint the Subcontract DAB. b) None.
20.6 Obtaining Subcontract DAB's Decision	In the case that the Subcontractor serves a notice of dissatisfaction with the subcontract DAB's decision, attempt to settle the dispute amicably.	Before the commencement of arbitration.	None.

THE OBLIGATIONS OF THE CONTRACTOR

Red Book Subcontract

THE OBLIGATIONS OF THE SUBCONTRACTOR

CLAUSE	OBLIGATIONS	TIME FRAME	SPECIFIC CONSEQUENCES OF NON-COMPLIANCE
GENERAL CONDITIONS			
1 Definitions and Interpretation			
1.5 Priority of Subcontract Documents	In the case that the Subcontractor becomes aware of an error or defect in a document prepared for the Subcontract Works, give notice to the Contractor.	Promptly.	None.
1.7 Joint and Several Liability Under the Subcontract	a) In the case of a joint venture, consortium or other unincorporated grouping of two or more persons, notify the Contractor of the leader. b) Not to alter the composition or legal status of the joint venture without the prior consent of the Contractor.	None.	None.
1.9 Subcontract Agreement	Enter into and execute a Subcontract Agreement.	Within 28 days after the Subcontractor receives the Contractor's Letter of Acceptance, unless agreed otherwise.	None.
2 The Main Contract			
2.1 Subcontractor's Knowledge of Main Contract	Give notice of any ambiguity or discrepancy discovered when reviewing the Subcontract and the Main Contract.	Promptly.	None.

CLAUSE	OBLIGATIONS	TIME FRAME	SPECIFIC CONSEQUENCES OF NON-COMPLIANCE
2.2 Compliance with Main Contract	a) In relation to the Subcontract Works, perform and assume all the obligations and liabilities of the Contractor under the Main Contract with the exception of the Clauses listed. b) Design (to the extent provided for by the Subcontract), execute and complete the Subcontract Works and remedy and defects.	a) None. b) In such good time so as not to cause any breach by the Contractor.	None.
2.3 Instructions and Determinations under Main Contract	a) Comply with all instructions and determinations of the Engineer that are notified as a Contractor's Instruction. b) If the Subcontractor receives any direct instructions from the Employer or the Engineer, inform the Contractor's Subcontractor Representative and supply a copy of any written instruction given.	a) None. b) Immediately.	None.
3 The Contractor			
3.1 Contractor's Instructions	a) Take instructions only from the Contractor's Subcontractor Representative. b) Comply with all instructions, given or confirmed in writing on any matter related to the Subcontract.	None.	None.
3.4 Employer's Claims in connection with the Main Contract	Provide all reasonable assistance to the Contractor in relation to Employer's claims.	None.	None.
3.5 Co-ordination of Main Works	a) Submit details of the arrangements and methods which the Subcontractor proposes to adopt for the execution of the Subcontract Works. b) Not to make any significant alteration to the arrangements and methods without the Contractor's prior consent.	a) Whenever required by a Contractor's Instruction. b) None.	None.

Red Book
Subcontract

THE OBLIGATIONS OF THE SUBCONTRACTOR (continued)

CLAUSE	OBLIGATIONS	TIME FRAME	SPECIFIC CONSEQUENCES OF NON-COMPLIANCE
4 The Subcontractor			
4.1 Subcontractor's General Obligations	a) Design (to the extent provided for by the Subcontract), execute and complete the Subcontract Works and remedy and defects in accordance with the Subcontract and with the Contractor's Instructions. b) Be responsible for any work designed by the Subcontractor. c) Provide all personnel, superintendence, labour, Subcontractor's Plant, Subcontractor's Equipment, Subcontractor's Documents and all other things required for the design, completion and the remedying of any defects. d) Be responsible for the adequacy, stability and safety of all the Subcontract Site Operations and methods of construction.	None.	None.
4.2 Subcontract Performance Security	Obtain a Subcontract Performance Security and deliver it to the Contractor.	Within 28 days of receiving the Contractor's Letter of Acceptance.	The Contractor is not obliged to make any payment to the Subcontractor (Sub-Clause 14.6).
4.3 Access to the Subcontract Works	Permit the Employer and the Contractor's Personnel to have full access to examine, inspect, measure and test the materials and workmanship and to check the progress of the Subcontract Works.	At all reasonable times.	None.

CLAUSE	OBLIGATIONS	TIME FRAME	SPECIFIC CONSEQUENCES OF NON-COMPLIANCE
5 Assignment of the Subcontract and Subcontracting			
5.1 Assignment of Subcontract	Not to assign the whole or any part of the Subcontract or any benefit or interest in or under the Subcontract without prior consent of the Contractor, except for instances listed in this Sub-Clause.	None.	None.
5.2 Subcontracting	a) Not to subcontract the whole or any part of the Subcontract without the prior consent of the Contractor. b) Give the Contractor notice of the intended date of the commencement of each subcontractor's work and of the commencement of that work on the Site. c) Ensure that each of the subcontracts include provisions which would entitle the Contractor to require the benefits of the Subcontractor's obligations under the Subcontract be assigned to the Contractor.	a) None. b) Not less than 14 days. c) None.	None.
6 Co-operation, Staff and Labour			
6.1 Co-operation under the Subcontract	a) Co-operate with and allow appropriate opportunities for carrying out work by the Employer's personnel and contractors, the Contractor, the Contractor's personnel, the Contractor's subcontractors and persons from public authorities. b) In the case that non-co-operation of the above affect the Subcontract work, notify the Contractor. c) In the case that non-co-operation of the above causes the Subcontractor to suffer delay and/or incur Cost, give notice to the Contractor.	a) None. b) Immediately. c) As soon as practicable and not later than 21 days after the Contractor became aware, or should have become aware of the event or circumstance (Sub-Clause 20.2).	a) None. b) None. c) Loss of entitlement to an extension to the subcontract Time for Completion and additional payment (Sub-Clause 20.2).

THE OBLIGATIONS OF THE SUBCONTRACTOR

Red Book
Subcontract

THE OBLIGATIONS OF THE SUBCONTRACTOR (continued)

CLAUSE	OBLIGATIONS	TIME FRAME	SPECIFIC CONSEQUENCES OF NON-COMPLIANCE
6.2 Persons in the Service of Others	Not to recruit, or attempt to recruit, staff and labour from amongst the Contractor's Personnel.	None.	None.
6.4 Subcontractor's Representative	a) Appoint the Subcontractor's Representative and give him all authority necessary to act on the Subcontractor's behalf under the Subcontract. b) Ensure that the whole of the time of the Subcontractor's Representative is given to directing the Subcontractor's performance of the Subcontract. c) In the case that the Subcontractor's Representative is to be temporarily absent, obtain the Contractor's consent and appoint a replacement person. d) Ensure that the Subcontractor's Representative receives instructions on behalf of the Subcontractor. e) Ensure that the Subcontractor's Representative is fluent in the langue for communications.	None.	None.
7 Equipment, Temporary Works, Other Facilities, Plant and Materials			
7.1 Subcontractor's Use of Equipment, Temporary Works and/or Other Facilities	a) Visually inspect any specified equipment, temporary works and/or other facilities that are made available to the Subcontractor. b) In the case of any shortage, defect or default, give notice to the Contractor. c) Not to remove from the Site any items of Employer's Equipment or Contractor's Equipment without the consent of the Contractor.	a) None. b) Promptly. c) None.	None.

CLAUSE	OBLIGATIONS	TIME FRAME	SPECIFIC CONSEQUENCES OF NON-COMPLIANCE
7.2 Free-Issue Materials	a) Visually inspect the specified free-issue materials. b) In the case of any shortage, defect or default, give notice to the Contractor.	a) None. b) Promptly.	None.
7.5 Subcontract Equipment and Subcontract Plant	a) Be responsible for all Subcontractor's Equipment. b) Not to remove from the Site any major items of Subcontractor's Equipment without the consent of the Contractor.	None.	None.
8 Commencement and Completion			
8.1 Commencement of Subcontract Works	a) Commence the execution of the Subcontract Works. b) Proceed with the Subcontract Works with due expedition and without delay.	a) As soon as is reasonably possible after the Subcontract Commencement Date. b) None.	None.
8.2 Subcontract Time for Completion	Complete the Subcontract Works.	Within the Subcontract Time for Completion or the extension of such.	The Contractor is entitled to deduct delay damages (Sub-Clause 8.7)
8.3 Subcontract Programme	a) Submit a detailed programme for the execution of the Subcontract Works. b) In the case that the Contractor issues a Contractor's Instruction to do so, submit a revised programme and supporting report describing revised methods which the Subcontractor proposes to adopt in order to expedite progress. c) Adopt revised methods to expedite progress.	a) Within 14-days of receipt of the Letter of Acceptance or the Contractor's programme, whichever is the later. d) None. e) None.	None.

THE OBLIGATIONS OF THE SUBCONTRACTOR

Red Book Subcontract

THE OBLIGATIONS OF THE SUBCONTRACTOR (continued)

CLAUSE	OBLIGATIONS	TIME FRAME	SPECIFIC CONSEQUENCES OF NON-COMPLIANCE
8.5 Subcontract Progress Reports	If required by the Contractor, submit monthly progress reports to the Contractor.	No later than 5 days before the due date for submission of the Contractor's progress report.	None.
8.6 Suspension of Subcontract Works by the Contractor	a) In the case of suspension of all or part of the Subcontract Works, comply with any obligations detailed under Main Contract Clauses 8.9, 8.10, 8.11 and 8.12. b) Not to suspend progress of part or all of the Subcontract Works unless and until required by a Contractor's instruction.	a) As appropriate. b) None.	a) As appropriate. b) None.
9 Tests on Completion			
9.1 Subcontract Tests on Completion	a) Give notice to the Contractor of the date after the Subcontractor will be ready to carry out any specified tests. b) In the case of failure to pass the tests, make good any defects and repeat the tests.	a) Within reasonable time. b) As soon as practicable.	None.
9.2 Main Contract Tests on Completion	In the case that the Subcontract specifies or makes reference to the Main Contract Tests on Completion, comply with Main Contract Clause 9.	As appropriate.	As appropriate.
10 Completion and Taking-Over the Subcontract Works			
10.1 Completion of Subcontract Works	a) Notify the Contractor when the Subcontract Works will be complete. b) In the case that the Contractor gives notice that completion has not been achieved, complete the work and issue a further notice.	a) Not earlier than 7 days before the Subcontract Works will be complete. b) None.	None.

CLAUSE	OBLIGATIONS	TIME FRAME	SPECIFIC CONSEQUENCES OF NON-COMPLIANCE
11 Defects Liability			
11.1 Subcontractor's Obligations after Taking-Over	Comply with the provisions of Main Contract Clauses 11.1, 11.4, 11.5, 11.6, 11.7 and 11.8.	As appropriate.	As appropriate.
12 Measurement and Evaluation			
12.1 Measurement of Subcontract Works	a) Supply the Contractor with any particulars requested by the Contractor and/or Engineer. b) In the case that the Subcontractor disagrees with the records, give notice to the Contractor.	a) None. b) Within 7 days of the date of the examination of the records.	a) None. b) The records shall be deemed to be accurate and to be accepted by the Subcontractor.
12.3 Evaluation under the Subcontract	Give effect to each agreement reached or evaluation made under this Sub-Clause.	Until revised under Clause 20 (*Claims and Disputes*).	None.
13 Subcontract Variations and Adjustments			
13.1 Variation of Subcontract Works	a) Execute and be bound by each Subcontract Variation. b) In the case that the Subcontractor cannot readily obtain the Subcontract Goods required for the Subcontract Variation, give notice with supporting particulars to the Contractor.	a) None. b) Promptly.	None.
13.3 Request for Proposal for Subcontract Variation	a) Respond in writing to a request for a proposal from the Contractor prior to instructing a Subcontract Variation. b) Not delay any work whilst awaiting a response in respect of a proposal.	a) As soon as practicable. b) None.	None.

Red Book Subcontract

THE OBLIGATIONS OF THE SUBCONTRACTOR (continued)

CLAUSE	OBLIGATIONS	TIME FRAME	SPECIFIC CONSEQUENCES OF NON-COMPLIANCE
13.4 Subcontract Adjustments for Changes in Legislation	In the case that the Subcontractor suffers delay and/or incurs Cost as a result of a change in the Laws, give notice to the Contractor.	As soon as practicable and not later than 21 days after the Contractor became aware, or should have become aware of the event or circumstance (Sub-Clause 20.2).	Loss of entitlement to an extension to the Subcontract Time for Completion and additional payment (Sub-Clause 20.2).
14 Subcontract Price and Payment			
14.1 The Subcontract Price	a) Pay all taxes, duties and fees required to be paid under the Subcontract. b) Submit to the Contractor a breakdown of each lump-sum price in the Subcontract.	a) None. b) Within 7 days of the Contractor's request.	None.
14.2 Subcontract Advance Payment	a) In the case that an advance payment is included in the Appendix to the Subcontractor's Offer, provide an advance payment guarantee. b) Ensure that the advance payment guarantee is valid and enforceable until the advance payment has been repaid.	None.	None.
14.3 Subcontractor's Monthly Statements	a) Prepare and submit monthly statements and supporting documents to the Contractor, showing the amount that the Subcontractor considers himself entitled. b) Include the items set out in Contract Clause 14.3, paragraphs (a) to (d), (f) and (g) in each monthly statement.	No later than 7 days before the due date for submission of the Contractor's Statement.	The Contractor has no obligation to make provision for the Subcontractor's amount within his next Statement (Sub-Clause 14.5).

CLAUSE	OBLIGATIONS	TIME FRAME	SPECIFIC CONSEQUENCES OF NON-COMPLIANCE
14.4 Subcontractor's Statement at Completion	Submit a statement showing the information relevant to the Subcontract as set out in sub-paragraphs (a) to (c) of Main Contract Clause 14.10.	No later than 7 days before the due date of submission of the Contractor's Statement at Completion, or no later that 28 days after the Subcontract Works have been taken over by the Contractor, whichever is applicable.	None.
14.8 Final Subcontract Payment	a) Prepare and submit a draft final statement stating the sum which, in the Subcontractor's opinion, is finally due. b) Submit further information that the Contractor may reasonably require.	a) No later than 28 days after expiry of the Subcontract Defects Notification Period. c) None.	None.
16 Suspension and Termination by the Subcontractor			
16.1 Subcontractor's Entitlement to Suspend Work	a) If the Subcontractor intends to suspend work (or reduce the rate of work) in the case of non-payment by the Contractor, give notice to the Contractor. b) In the case that the Subcontractor receives payment, resume normal working. c) In the case that the Subcontractor suffers delay and/or incurs Cost as a result of to suspending work (or reducing the rate of work), give notice to the Contractor.	a) Not less than 21 days before it is intended to suspend work (or reduce the rate of work). b) As soon as is reasonably practicable. c) Within 21 days (Sub-Clause 20.2).	a) None. b) None. c) Loss of entitlement to an extension to the Subcontract Time for Completion and additional payment (Sub-Clause 20.2).
16.2 Termination by Subcontractor	In the case that the Subcontractor intends to terminate the Subcontract, give notice to the Contractor.	Not less than 14-days or immediately in the case of the Contractor or the Employer becoming insolvent or bankrupt.	None.

**Red Book
Subcontract**

THE OBLIGATIONS OF THE SUBCONTRACTOR (continued)

CLAUSE	OBLIGATIONS	TIME FRAME	SPECIFIC CONSEQUENCES OF NON-COMPLIANCE
17 Risk and Indemnities			
17.1 Subcontractor's Risks and Indemnities	a) Have full responsibility for the care of the Subcontract Works and Goods. b) Take responsibility for any work which is outstanding. c) Be liable for any loss or damage caused by the Subcontractor. d) Be liable for any loss or damage arising from a previous event for which the Subcontractor was liable. e) In the case of loss or damage, rectify the loss or damage.	a) From the Subcontract Commencement Date until the Subcontract Works have been taken over. b)–d) After responsibility for the Subcontract Works has passed to the Employer or the Contractor. e) Without delay.	None.
18 Subcontract Insurances			
18.1 Subcontractor's Obligation to Insure	a) Effect and maintain the insurances in accordance with the details set out in Annexe E. b) Insure the Subcontractor's Personnel in accordance with Main Contract Clause 18.	From the Subcontract Commencement Date until the Subcontract Works have been taken over.	Contractor shall effect the insurance, pay the premium and be entitled to reimbursement of the cost, plus expenses incurred. (Sub-Clause 18.3).
18.2 Insurance arranged by the Contractor and/or the Employer	In the case that the Subcontractor discovers any inadequacy in the insurances arranged by the Contractor and/or the Employer, or duplication, notify the Contractor.	Immediately.	None.

CLAUSE	OBLIGATIONS	TIME FRAME	SPECIFIC CONSEQUENCES OF NON-COMPLIANCE
18.3 Evidence of Insurance and Failure to Insure	If so requested by the Contractor, provide evidence of the insurance and payment of the current premium.	Promptly.	Contractor shall effect the insurance, pay the premium and be entitled to reimbursement of the cost, plus expenses incurred.
20 Notices, Subcontractor's Claims and Disputes			
20.1 Notices	a) Whenever the Contractor is required to provide notice or other information, or to keep contemporary records, to the extent that these terms apply to the Subcontract Works, give notice or information and keep contemporary records that will enable the Contractor to comply with the terms of the Main Contract. b) Give notice to the Contractor of any delay event which has occurred or specific probable future event(s) or circumstance(s) which may adversely affect the Contractor's activities or delay the execution of the Subcontract Works and/or the Main Works. c) Give notice to the Contractor of any event which has occurred or specific probable future event(s) or circumstance(s) which may increase the Subcontract Price and/or the Contract Price.	a) In good time so as to enable the Contractor to comply with the terms of the Main Contract. b) Immediately. c) Immediately.	1. Loss of entitlement to an extension to the Subcontract Time for Completion and additional payment (Sub-Clause 20.2). 2. If the Contractor is prevented from recovering any sum from the Employer, the Contractor shall be entitled to recover this sum from the Subcontractor (Sub-Clause 20.3).

Red Book Subcontract

THE OBLIGATIONS OF THE SUBCONTRACTOR (continued)

CLAUSE	OBLIGATIONS	TIME FRAME	SPECIFIC CONSEQUENCES OF NON-COMPLIANCE
20.2 Subcontractor's Claims	a) In the case that the Subcontractor considers himself entitled to any extension of the Subcontract Time for Completion and/or additional payment, comply with the obligations of the Contractor set out in Main Contract Clause 20.1. b) Give effect to each agreement reached or decision made unless and until revised.	Time frames included in Main Contract Clause 20.1 are adjusted as follows: 1. The period of notice is 21 days. 2. The period for submission of a fully detailed claim is 35 days.	1. In the case of failure to give notice of a claim, the Subcontractor shall lose entitlement to an extension to the Subcontract Time for Completion and additional payment. 2. If the Contractor is prevented from recovering any sum from the Employer, the Contractor shall be entitled to recover this sum from the Subcontractor (Sub-Clause 20.3).
20.4 Subcontract Disputes	Afford the Contractor all information and assistance that may be required to enable the Contractor to pursue any dispute, which includes a Subcontract dispute under the Main Contract.	In good time.	None.
20.5 Appointment of the Subcontract DAB	a) Jointly, with the Contractor, appoint the DAB. b) Not to act alone in the termination of any member of the Dispute Adjudication Board.	a) Within 42 days after the date of a Notice of Dispute. b) None.	a) The President of FIDIC or his appointee may appoint the Subcontract DAB. b) None.
20.6 Obtaining Subcontract DAB's Decision	In the case that the Contractor serves a notice of dissatisfaction with the subcontract DAB's decision, attempt to settle the dispute amicably.	Before the commencement of arbitration.	None.

Chapter 4

The Yellow Book

Conditions of Contract for Plant and Design-Build for Electrical and Mechanical Plant, and for Building and Engineering Works, Designed by the Contractor, First Edition 1999

The FIDIC Contracts: Obligations of the Parties, First Edition. Andy Hewitt.
© 2014 John Wiley & Sons, Ltd. Published 2014 by John Wiley & Sons, Ltd.

THE OBLIGATIONS OF THE EMPLOYER

CLAUSE	OBLIGATIONS	TIME FRAME	SPECIFIC CONSEQUENCES OF NON-COMPLIANCE
GENERAL CONDITIONS			
1 General Provisions			
1.6 Contract Agreement	Enter into a Contract Agreement with the Contractor.	Within 28 days after the Contractor receives the Letter of Acceptance, unless agreed otherwise.	None.
1.13 Compliance with Laws	Obtain the planning, zoning or similar permission for the Permanent Works, and any other permissions described in the Employers' Requirements as having been (or being) obtained by the Employer.	None.	None.
2 The Employer			
2.1 Right of Access to the Site	a) Give the Contractor right of access to, and possession of, all parts of the Site. b) Give the Contractor possession of any foundation, structure, plant or means of access if required.	a) Within the time (or times) stated in the Appendix to Tender, or if not stated, to enable the Contractor to proceed in accordance with the programme submitted under Sub-Clause 8.3 [Programme]. b) In the time and manner stated in the Employer's Requirements.	Contractor shall be entitled an extension of time and payment of Cost plus reasonable profit.
2.2 Permits, Licences or Approvals	Provide reasonable assistance to the Contractor at the request of the Contractor: a) By obtaining copies of the Laws of the Country which are relevant but not readily available and b) For the Contractor's applications for any permits, licences or approvals required by the Laws of the Country.	None.	None.

THE OBLIGATIONS OF THE EMPLOYER (continued)

CLAUSE	OBLIGATIONS	TIME FRAME	SPECIFIC CONSEQUENCES OF NON-COMPLIANCE
2.3 Employer's Personnel	Ensure that Employer's Personnel and Employer's other contractors cooperate with the Contractor and take actions similar to those which the Contractor is required to take under Sub-Clause 4.8 [Safety Procedures] and under Sub-Clause 4.18 [Protection of the Environment].	None.	None.
2.4 Employer's Financial Arrangements	Submit reasonable evidence that financial arrangements have been made and are being maintained which will enable the Employer to pay the Contract Price.	Within 28 days after receiving any request from the Contractor.	Contractor entitled to suspend work, reduce the rate of work and to an extension of time and additional payment as a result of such actions (Sub-Clause 16.1).
2.5 Employer's Claims	Give notice and particulars to the Contractor if the Employer considers himself to be entitled to any payment under any Clause of these Conditions or otherwise in connection with the Contract and/or to any extension of the Defects Notification Period (this obligation may be carried out by the Engineer).	As soon as practicable after becoming aware of the event or circumstances giving rise to the claim.	None.
3 The Engineer			
3.1 Engineer's Duties and Authority	a) Appoint the Engineer to carry out the duties assigned to him in the Contract. b) Not to impose further constraints on the Engineer's authority, except as agreed with the Contractor.	None.	None.

CLAUSE	OBLIGATIONS	TIME FRAME	SPECIFIC CONSEQUENCES OF NON-COMPLIANCE
3.4 Replacement of the Engineer	a) Give notice to the Contractor of the name, address and relevant experience of the intended replacement Engineer. b) Not to replace the Engineer with a person against whom the Contractor raises reasonable objection by notice to the Employer, with supporting particulars.	a) Not less than 42 days before the intended date of replacement. b) None.	None.
4 The Contractor			
4.2 Performance Security	a) Cooperate with the Contractor to agree the entity, country (or other jurisdiction) for the issue of the Performance Security. b) Cooperate with the Contractor to agree the form of Performance Security if not in the form annexed to the Particular Conditions. c) Not to make a claim under the Performance Security, except for amounts to which the Employer is entitled under the Contract (as listed). d) Return the Performance Security to the Contractor.	a) None. b) None. c) None. d) Within 21 days after receiving a copy of the Performance Certificate.	None.
4.10 Site Data	a) Make available to the Contractor all relevant data in the Employer's possession on sub-surface and hydrological conditions at the Site, including environmental aspects. b) Make available to the Contractor all such data which come into the Employer's possession after the Base Date.	a) Prior to the Base Date. b) None.	None.

THE OBLIGATIONS OF THE EMPLOYER

Yellow Book

THE OBLIGATIONS OF THE EMPLOYER (continued)

CLAUSE	OBLIGATIONS	TIME FRAME	SPECIFIC CONSEQUENCES OF NON-COMPLIANCE
4.20 Employer's Equipment and Free-Issue Material	a) Make the Employer's Equipment (if any) available for the use of the Contractor in the execution of the Works in accordance with the details, arrangements and prices stated in the Employer's Requirements. b) Supply, free of charge, the 'free-issue materials' (if any) in accordance with the details stated in the Employer's Requirements. c) Rectify the notified shortage, defect or default in the free-issue materials.	a) As specified in the Contract. b) As specified in the Contract. c) Immediately.	None.
4.24 Fossils	Take possession and care of fossils, coins, articles of value or antiquity and structures and other remains, or items of geological or archaeological interest found on the Site.	None.	None.
10 Employer's Taking Over			
10.1 Taking Over of the Works and Sections	Take over the Works.	When the Works have been completed in accordance with the Contract and a Taking-Over Certificate has been issued.	None.
10.2 Taking Over of Parts of the Works	Not to use any part of the Works (other than as a temporary measure which is either specified in the Contract or agreed by both Parties) unless and until the Engineer has issued a Taking-Over Certificate for this part.	None.	If the Contractor incurs a Cost as a result of the Employer taking over and/or using a part of the Works, the Contractor shall be entitled to payment of any such Cost plus reasonable profit.

CLAUSE	OBLIGATIONS	TIME FRAME	SPECIFIC CONSEQUENCES OF NON-COMPLIANCE
11 Defects Liability			
11.2 Cost of Remedying Defects	Notify the Contractor (or ensure that notice is given by others) of any work to be remedied if due to any cause outside the provisions of the Contract.	Promptly.	None.
11.4 Failure to Remedy Defects	In the case of failure by the Contractor to remedy any defect or damage, notify the Contractor (or ensure that notice is given by others) of the date by which the defect or damage is to be remedied.	Within reasonable time.	None.
12 Tests after Completion			
12.1 Procedure for Tests after Completion	a) If Tests after Completion are specified, provide all electricity, equipment, fuel, instruments, labour, materials and suitably qualified and experienced staff. b) Carry out the Tests after Completion in accordance with the manuals and guidance provided by the Contractor. c) Give the Contractor notice of the date when the Tests after Completion will be carried out. d) Compile and evaluate the results of the Tests after Completion jointly with the Contractor.	a) None. b) None. c) 21 days before the date of the Tests. d) None.	None.
14 Contract Price and Payment			
14.2 Advance Payment	a) Make the advance payment, as an interest-free loan for mobilisation. b) Cooperate with the Contractor to approve the form of the advance payment guarantee.	a) As stated in the Appendix to Tender and when the Contractor submits a guarantee in accordance with this Sub-Clause. b) None.	None.

THE OBLIGATIONS OF THE EMPLOYER

Yellow Book

THE OBLIGATIONS OF THE EMPLOYER (continued)

CLAUSE	OBLIGATIONS	TIME FRAME	SPECIFIC CONSEQUENCES OF NON-COMPLIANCE
14.5 Plant and Material intended for the Works	Cooperate with the Contractor to approve the form of a bank guarantee for shipped Plant and Materials.	None.	None.
14.7 Payment	a) Pay the first instalment of the advance payment. b) Pay the amount certified in each Interim Payment Certificate. c) Pay the amount certified in the Final Payment Certificate.	a) Within 42 days after issuing the Letter of Acceptance or within 21 days after receiving the documents in accordance with Sub-Clause 4.2 [Performance Security] and Sub-Clause 14.2 [Advance Payment], whichever is later. b) Within 56 days after the Engineer receives the Statement and supporting documents. c) Within 56 days after the Employer receives the Final Payment Certificate	1) Contractor entitled to suspend work, reduce the rate of work and to an extension of time and additional payment as a result of such actions (Sub-Clause 16.1). 2) Payment of financing charges to the Contractor.
14.9 Payment of Retention Money	a) Pay the Contractor the first half of the Retention Money. b) Pay the Contractor the outstanding balance of the Retention Money.	a) When a Taking-Over Certificate has been issued and when the Engineer has certified payment. b) Promptly after the latest of the expiry of the Defects Notification Periods and when the Engineer has certified payment.	The Contractor is entitled to receive financing charges (Sub-Clause 14.7).
14.15 Currencies of Payment	Pay the Contractor in the currency or currencies named in the Appendix to Tender.	None.	None.

CLAUSE	OBLIGATIONS	TIME FRAME	SPECIFIC CONSEQUENCES OF NON-COMPLIANCE
15 Termination by Employer			
15.2 Termination by Employer	a) Give notice of intention to terminate the Contract. b) Give notice of release of the Contractor's Equipment and Temporary Works.	a) 14 days prior to termination date, or immediately in the case of the Contractor becoming bankrupt or gives or offers bribes or gratuities (or similar as defined in the clause). b) On completion of the Works.	None.
15.4 Payment after Termination	Pay the balance due to the Contractor after recovering any losses, damages and extra costs.	None.	None.
15.5 Employer's Entitlement to Termination	a) Give notice of intention to terminate the Contract. b) Return the Performance Security. c) Not to terminate the Contract in order to execute the Works himself or to arrange for the Works to be executed by another contractor.	a) 28 days prior to the termination date. b) None. c) None.	None.
16 Suspension and Termination by Contractor			
16.4 Payment on Termination	a) Return the Performance Security to the Contractor. b) Pay the Contractor in accordance with Sub-Clause 19.6 *[Optional Termination, Payment and Release]*. c) Pay to the Contractor the amount of any loss of profit or other loss or damage sustained by the Contractor as a result of this termination.	Promptly.	None.

Yellow Book

THE OBLIGATIONS OF THE EMPLOYER (continued)

CLAUSE	OBLIGATIONS	TIME FRAME	SPECIFIC CONSEQUENCES OF NON-COMPLIANCE
17 Risk and Responsibility			
17.1 Indemnities	Indemnify and hold harmless the Contractor, the Contractor's Personnel and their respective agents against and from all claims, damages, losses and expenses in respect of bodily injury, disease or death which is attributable to any negligence, willful act or breach of the Contract by the Employer, the Employer's Personnel or agents.	None.	None.
17.5 Intellectual and Industrial Property Rights	a) Give notice of any claim under this clause. b) Indemnify and hold the Contractor harmless against and from any claim, which is or was an unavoidable result of the Contractor's compliance with the Contract, or as a result of any Works being used by the Employer. c) If requested by the Contractor, assist in contesting the claim. d) Not to make any admission, which might be prejudicial to the Contractor.	a) Within 28 days of receiving a claim. b) None. c) None. d) None.	a) Waiver of right to indemnity. b) None. c) None. d) None.

CLAUSE	OBLIGATIONS	TIME FRAME	SPECIFIC CONSEQUENCES OF NON-COMPLIANCE
18 Insurance			
18.1 General Requirements for Insurances	a) Cooperate with the Contractor to approve the terms of insurances. b) Effect and maintain the insurances in terms consistent with the details annexed to the Particular Conditions, wherever the Employer is the insuring Party. c) Submit evidence that the insurance has been effected, provide copies of the policies and submit evidence of payment. d) Inform the insurers of any relevant changes to the execution of the Works and ensure that insurance is maintained. e) Not to make any material alteration to the terms of any insurance without approval of the Contractor.	a) None. b) None. c) Within the time frames stipulated in the Appendix to Tender. d) As appropriate. e) None.	a) None. b) None. c) The Contractor may effect the insurance and the Contract Price shall be adjusted. d) None. e) None.
19 Force Majeure			
19.2 Notice of Force Majeure	Give notice to the Contractor in the case that the Employer is, or will be prevented from performing the Employer's obligations by Force Majeure.	Within 14 days of becoming aware of the event.	None.
19.3 Duty to Minimise Delay	a) Use all reasonable endeavours to minimise any delay in the performance of the Contract as a result of Force Majeure. b) Give notice when the effects of the Force Majeure cease.	None.	None.

THE OBLIGATIONS OF THE EMPLOYER (continued)

CLAUSE	OBLIGATIONS	TIME FRAME	SPECIFIC CONSEQUENCES OF NON-COMPLIANCE
20 Claims, Disputes and Arbitration			
20.2 Appointment of the Dispute Adjudication Board	a) Jointly appoint the DAB. b) Not to consult the DAB without the agreement of the Contractor. c) Not to act alone in the termination of any member of the Dispute Adjudication Board.	a) Within 28 days after a Party gives notice of its intention to refer a dispute to a DAB. b) None. c) None.	a) The appointing entity or official named in the Appendix to Tender shall appoint (Sub-Clause 20.3). b) None. c) None.
20.4 Obtaining Dispute Adjudication Board's Decision	a) Make available to the DAB additional information, access to the Site, and appropriate facilities as the DAB may require. b) Give effect to a DAB decision unless and until it is revised in an amicable settlement or an arbitral award.	a) Promptly. b) None.	a) None. b) The matter may be referred to arbitration (Sub-Clause 20.7).
20.5 Amicable Settlement	In the case of a notice of dissatisfaction being issued, attempt to settle the dispute amicably.	Within 56 days of the notice.	Arbitration may be commenced.

CLAUSE	OBLIGATIONS	TIME FRAME	SPECIFIC CONSEQUENCES OF NON-COMPLIANCE
GENERAL CONDITIONS OF DISPUTE ADJUDICATION AGREEMENT			
2 General Provisions	Give notice to the Dispute Adjudication Board Member that the Dispute Adjudication Agreement has taken effect.	When all parties have signed the dispute adjudication agreement.	None.
5 General Obligations of the Employer and the Contractor	Not to request advice from, or consultation with a DAB Member regarding the Contract otherwise than in the normal course of the DAB's activities.	None.	None.
6 Payment	a) Pay one half of the DAB fees to the Contractor. b) In the Case of the Contractor failing to pay the DAB Member, pay the due fees.	a) Within the monthly payments to the Contractor. b) None.	The DAB Member may suspend services or resign the appointment.
Annex – Procedural Rules			
1.	a) Jointly with the Contractor, furnish to each DAB member one copy of all documents which the DAB may request. b) Copy all communications with the DAB to the Contractor.	None.	None.

THE OBLIGATIONS OF THE CONTRACTOR

CLAUSE	OBLIGATIONS	TIME FRAME	SPECIFIC CONSEQUENCES OF NON-COMPLIANCE
GENERAL CONDITIONS			
1 Definitions			
1.6 Contract Agreement	Enter into a Contract Agreement.	Within 28 days after the Contractor receives the Letter of Acceptance, unless agreed otherwise.	None.
1.8 Care and Supply of Documents	a) Supply to the Engineer six copies of each of the Contractor's Documents. b) Keep, on the Site, a copy of the Contract, publications named in the Employer's Requirements, the Contractor's Documents and Variations and other communications given under the Contract. c) In the case of an error in a document, give notice to the Employer.	a) None. b) None. c) Promptly.	None.
1.9 Errors in the Employer's Requirements	a) Give notice to the Employer, of errors or defects of a technical nature in a document which was prepared for use in executing the Works. b) Give notice to the Engineer if the Contractor suffers delay and/or incurs Cost as a result of an error in the Employer's Requirements.	a) Promptly. b) As soon as practicable and no later than 28 days after the Contractor became aware, or should have become aware of the event or circumstance.	a) None. b) Loss of entitlement to an extension to the Time for Completion and additional payment (Sub-Clause 20.1).
1.11 Contractor's Use of Employer's Documents	Not to, without consent, copy, use or communicate the Employer's Requirements and other documents of the Employer to a third party.	None.	None.

CLAUSE	OBLIGATIONS	TIME FRAME	SPECIFIC CONSEQUENCES OF NON-COMPLIANCE
1.12 Confidential Details	Disclose all such confidential and other information as the Engineer may reasonably require in order to verify the Contractor's compliance with the Contract.	None.	None.
1.13 Compliance with Laws	a) Comply with applicable Laws. b) Give all notices, pay all taxes, duties and fees, and obtain all permits, licences and approvals, as required by the Laws.	None.	None.
1.14 Joint and Several Liability	a) In the case of a joint venture, consortium or other unincorporated grouping of two or more persons, notify the Employer of the leader. b) Not to alter the composition of the joint venture or legal status without the prior consent of the Employer.	None.	None.
2 The Employer			
2.1 Right of Access to the Site	Give notice to the Engineer if the Contractor suffers delay and/or incurs Cost as a result of failure to give right of access and possession of the Site.	As soon as practicable and not later than 28 days after the Contractor became aware, or should have become aware of the event or circumstance (Sub-Clause 20.1).	Loss of entitlement to an extension to the Time for Completion and additional payment (Sub-Clause 20.1).
3 The Engineer			
3.3 Instructions of the Engineer	a) Only take instructions from the Engineer, or from an assistant to whom the appropriate authority has been delegated. b) Comply with the instructions given by the Engineer or delegated assistant on any matter related to the Contract.	None.	None.

Yellow Book

THE OBLIGATIONS OF THE CONTRACTOR (continued)

CLAUSE	OBLIGATIONS	TIME FRAME	SPECIFIC CONSEQUENCES OF NON-COMPLIANCE
4 The Contractor			
4.1 Contractor's General Obligations	a) Design, execute and complete the Works in accordance with the Contract and remedy any defects in the Works. b) Provide the required Plant and Contractor's Documents specified in the Contract and all Contractor's Personnel, Goods, consumables and other things and services, whether of a temporary or permanent nature. c) Be responsible for the adequacy, stability and safety of all Site operations and of all methods of construction. d) Submit details of the arrangements and methods proposed for the execution of the Works. e) Not to make significant alteration to the arrangements and methods without previous notice to the Engineer.	a) None. b) None. c) None. d) Whenever required by the Engineer. e) None.	None.
4.2 Performance Security	a) Obtain a Performance Security for proper performance and deliver to the Employer. b) Ensure that the Performance Security is valid and enforceable until the Contractor has executed and completed the Works and remedied any defects. c) Extend the validity of the Performance Security until the Works have been completed and any defects have been remedied.	a) Within 28 days after receiving the Letter of Acceptance. b) None. c) As required.	a) None. b) None. c) Employer may claim the full amount of the Performance Security.

CLAUSE	OBLIGATIONS	TIME FRAME	SPECIFIC CONSEQUENCES OF NON-COMPLIANCE
4.3 Contractor's Representative	a) Appoint the Contractor's Representative and give him all authority necessary to act on the Contractor's behalf under the Contract. b) Submit to the Engineer for consent, the name and particulars of the person the Contractor proposes to appoint. c) If consent is withheld or subsequently revoked, or if the appointed person fails to act, submit the name and particulars of another suitable person for such appointment. d) Not, without the prior consent of the Engineer, revoke the appointment of the Contractor's Representative or appoint a replacement. e) If the Contractor's Representative is to be temporarily absent from the Site, appoint a suitable replacement and notify the Engineer.	a) None. b) Prior to the Commencement Date. c) None. d) None. e) None.	None.
4.4 Subcontractors	a) Not to subcontract the whole of the Works. b) Be responsible for the acts or defaults of any Subcontractor, his agents or employees. c) Obtain prior consent of the Engineer to proposed Subcontractors not named in the Contract. d) Give the Engineer notice of the intended date of the commencement of each Subcontractor's work and of the commencement of such work on the Site.	a) None. b) None. c) None. d) Not less than 28 days.	None.

Yellow Book

THE OBLIGATIONS OF THE CONTRACTOR (continued)

CLAUSE	OBLIGATIONS	TIME FRAME	SPECIFIC CONSEQUENCES OF NON-COMPLIANCE
4.6 Co-operation	a) Allow appropriate opportunities for carrying out work to the Employer's Personnel, any other contractors employed by the Employer and the personnel of any legally constituted public authorities. b) Be responsible for the Contractor's construction activities and co-ordinate the Contractor's own activities with those of other contractors to the extent specified in the Employer's Requirements. c) Submit such documents which require the Employer to give to the Contractor possession of any foundation, structure, plant or means of access.	a) None. b) None. c) In the time and manner stated in the Employer's Requirements.	None.
4.7 Setting Out	a) Set out the Works in relation to original points, lines and levels of reference specified in the Contract or notified by the Engineer. b) Give notice to the Engineer if the Contractor suffers delay and/or incurs Cost as a result of error in the items of reference.	a) None. b) As soon as practicable and not later than 28 days after the Contractor became aware, or should have become aware, of the event or circumstance (Sub-Clause 20.1).	a) None. b) Loss of entitlement to an extension to the Time for Completion and additional payment (Sub-Clause 20.1).
4.8 Safety Procedures	a) Comply with all applicable safety regulations. b) Take care for the safety of all persons entitled to be on the Site. c) Use reasonable efforts to keep the Site and Works clear of unnecessary obstruction. d) Provide fencing, lighting, guarding and watching of the Works. e) Provide any Temporary Works, which may be necessary for the use and protection of the public and the owners and occupiers of adjacent land.	None.	None.

CLAUSE	OBLIGATIONS	TIME FRAME	SPECIFIC CONSEQUENCES OF NON-COMPLIANCE
4.9 Quality Assurance	Institute a quality assurance system and submit details to the Engineer.	Before each design and execution stage is commenced.	None.
4.12 Unforeseeable Physical Conditions	a) Give notice of adverse physical conditions. b) Continue executing the Works, using such proper and reasonable measures as are appropriate for the physical conditions. c) Comply with any instructions which the Engineer may give. d) Give a further notice, if the Contractor suffers delay and/or incurs Cost due to unforeseen physical conditions.	a) As soon as practicable. b) None. c) None. d) As soon as practicable and not later than 28 days after the Contractor became aware, or should have become aware of the event or circumstance (Sub-Clause 20.1).	a) None. b) None. c) None. d) Loss of entitlement to an extension to the Time for Completion and additional payment (Sub-Clause 20.1).
4.13 Rights of Way and Facilities	a) Bear all costs and charges for special and/or temporary rights-of-way. b) Obtain any additional facilities outside the Site which the Contractor may require for the purposes of the Works.	None.	None.
4.14 Avoidance of Interference	Not to interfere with the convenience of the public, or the access to and use and occupation of all roads and footpaths.	None.	None.
4.15 Access Route	a) Use reasonable efforts to prevent any road or bridge from being damaged. b) Be responsible for any maintenance which may be required for the use of access routes. c) Provide all necessary signs or directions along access routes. d) Obtain any permission which may be required from the relevant authorities for use of routes, signs and directions.	None.	None.

THE OBLIGATIONS OF THE CONTRACTOR

Yellow Book

THE OBLIGATIONS OF THE CONTRACTOR (continued)

CLAUSE	OBLIGATIONS	TIME FRAME	SPECIFIC CONSEQUENCES OF NON-COMPLIANCE
4.16 Transport of Goods	a) Give the Engineer notice of the date on which any Plant or a major item of other Goods will be delivered to the Site. b) Be responsible for packing, loading, transporting, receiving, unloading, storing and protecting all Goods and other things required for the Works.	a) 21 days before delivery. b) None.	None.
4.17 Contractor's Equipment	a) Be responsible for all Contractor's Equipment. b) Not to remove from the Site any major items of Contractor's Equipment without the consent of the Engineer.	None.	None.
4.18 Protection of the Environment	a) Take all reasonable steps to protect the environment and to limit damage and nuisance to people and property. b) Ensure that emissions, surface discharges and effluent shall not exceed the values indicated in the Employer's Requirements and shall not exceed the values prescribed by applicable Laws.	None.	None.
4.19 Electricity, Water and Gas	a) Be responsible for the provision of all power, water and other services. b) Provide any apparatus necessary for use of services as may be available on the Site and for measuring the quantities consumed. c) Pay the Employer for the use of services available on the Site.	None.	None.

CLAUSE	OBLIGATIONS	TIME FRAME	SPECIFIC CONSEQUENCES OF NON-COMPLIANCE
4.20 Employer's Equipment and Free-Issue Materials	a) Be responsible for the Employers' Equipment when used by the Contractor. b) Pay the Employer for the use of the Employers' Equipment. c) Inspect the free-issue materials. d) Give notice of any shortage, defect or default in the free-issue materials.	a) None. b) None. c) None. d) Promptly.	None.
4.21 Progress Reports	Prepare and submit monthly progress reports.	Monthly, within 7-days of the period to which the report relates.	None.
4.22 Security of the Site	Keep unauthorised persons off the Site.	None.	None.
4.23 Contractor's Operations on Site	a) Confine operations to the Site and to any additional areas agreed as working areas. b) Take all necessary precautions to keep Contractor's Equipment and Contractor's Personnel within the Site and any agreed working areas. c) Keep the Site free from all unnecessary obstruction. d) Store or dispose of any Contractor's Equipment or surplus materials. e) Clear away and remove from the Site any wreckage, rubbish and Temporary Works. f) Leave the Site and the Works in a clean and safe condition.	None.	None.

Yellow Book

THE OBLIGATIONS OF THE CONTRACTOR (continued)

CLAUSE	OBLIGATIONS	TIME FRAME	SPECIFIC CONSEQUENCES OF NON-COMPLIANCE
4.24 Fossils	a) Take reasonable precautions to prevent Contractor's Personnel or other persons from removing or damaging fossils, coins, articles of value or antiquity, structures and other remains or items of geological or archaeological interest. b) Give notice of the finding of such items. c) Give further notice if the Contractor suffers delay and/or incurs Cost as a result of such items.	a) None. b) Upon discovery. c) As soon as practicable and not later than 28 days after the Contractor became aware, or should have become aware of the event or circumstance (Sub-Clause 20.1).	a) None. b) None. c) Loss of entitlement to an extension to the Time for Completion and additional payment (Sub-Clause 20.1).
5 Design			
5.1 General Design Obligations	a) Carry out and be responsible for the design of the Works. b) Use designers who are suitably qualified. c) Comply with the criteria stated in the Employer's Requirements. d) Submit to the Engineer for consent, the name and particulars of each proposed designer and design Subcontractor. e) Scrutinise the Employer's Requirements and setting out references. f) Give notice to the Engineer of any error, fault or other defect found in the Employer's Requirements or other references.	a) None. b) None. c) None. d) None. e) None. f) Within the period stated in the Appendix to Tender.	None.

CLAUSE	OBLIGATIONS	TIME FRAME	SPECIFIC CONSEQUENCES OF NON-COMPLIANCE
5.2 Contractor's Documents	a) Prepare all Contractors' Documents. b) Prepare all other documents necessary to instruct the Contractor's Personnel. c) In the case that the Employer's Requirements require it, submit the Contractor's Documents to the Engineer for review and/or approval. d) Give notice of the submission of Contractor's Documents to the Engineer for review and/or approval. e) In the case that the Engineer gives notice that a Contractor's Document does not comply with the Contract, rectify and resubmit the document for review and/or approval. f) In the case that Engineer's approval is required, do not commence execution of such part of the Works until the Engineer has approved the Contractor's Document or the approval period has expired. g) Execute parts of the Works in accordance with the reviewed and/or approved Contractor's Documents. h) Give notice to the Engineer in the case that the Contractor wishes to modify any design or document which has previously been submitted for review and/or approval. i) Submit revised documents to the Engineer in accordance with the Sub-Clause 5.2 procedure. j) In the case that the Engineer instructs that further Contractor's Documents are required, prepare the documents.	a) None. b) None. c) None. d) None. e) None. f) None. g) None. h) Immediately. i) None. j) Promptly.	None.

THE OBLIGATIONS OF THE CONTRACTOR (continued)

CLAUSE	OBLIGATIONS	TIME FRAME	SPECIFIC CONSEQUENCES OF NON-COMPLIANCE
5.3 Contractor's Undertaking	Design, provide the Contractor's Documents, execute and provide the completed Works in accordance with the Laws and the documents forming the Contract, as altered or modified by Variations.	None.	None.
5.4 Technical Standards and Regulations	a) Design, provide the Contractor's Documents, execute and provide the completed Works in accordance with the Country's technical standards, building, construction and environmental Laws, Laws applicable to the product being produced from the Works and other standards specified in the Employer's Requirements. b) Give notice to the Engineer in the case that new or applicable standards come into force after the Base Date and if appropriate submit proposals for compliance.	None.	None.
5.5 Training	Carry out the training of Employer's Personnel in the operation and maintenance of the Works.	None.	None.
5.6 As-Built Documents	a) Prepare and keep up to date, a complete set of as-built records of the execution of the Works. b) Provide two copies of the as-built records to the Engineer. c) Supply and submit to the Engineer for review, as-built drawings of the Works. d) Obtain the consent of the Engineer as to the size, referencing system and other relevant details of the as-built drawings. e) Supply to the Engineer the specified numbers of as-built drawings.	a) None. b) Prior to the commencement of the Tests on Completion. c) None. d) None. e) Prior to the issue of any Taking-Over Certificate.	a) None. b) None. c) None. d) None. e) The Works shall not be considered to be completed for the purposes of taking-over.

CLAUSE	OBLIGATIONS	TIME FRAME	SPECIFIC CONSEQUENCES OF NON-COMPLIANCE
5.7 Operation and Maintenance Manuals	a) Supply to the Engineer provisional operation and maintenance manuals. b) Supply to the Engineer final operation and maintenance manuals.	a) Prior to the commencement of the Tests on Completion. b) Prior to taking-over.	a) None. b) The Works shall not be considered to be completed for the purposes of taking-over.
5.8 Design Error	Correct any errors, omissions, ambiguities, inconsistencies, inadequacies or other defects that are found in the Contractor's Documents.	None.	None.
6 Staff and Labour			
6.1 Engagement of Staff and Labour	Make arrangements for the engagement of all staff and labour, local or otherwise and for their payment, housing, feeding and transport.	None.	None.
6.2 Rates of Wages and Conditions of Labour	Pay rates of wages and observe conditions of labour which are not lower than those established for the trade or industry where the work is carried out.	None.	None.
6.3 Persons in the Service of the Employer	Not to recruit, or attempt to recruit, staff and labour from amongst the Employer's personnel.	None.	None.
6.4 Labour Laws	a) Comply with all the relevant labour Laws. b) Require employees to obey all applicable Laws.	None.	None.
6.5 Working Hours	Obtain the consent of the Engineer if working outside the normal working hours.	None.	None.

THE OBLIGATIONS OF THE CONTRACTOR (continued)

CLAUSE	OBLIGATIONS	TIME FRAME	SPECIFIC CONSEQUENCES OF NON-COMPLIANCE
6.6 Facilities for Staff and Labour	a) Provide and maintain all necessary accommodation and welfare facilities for the Contractor's Personnel. b) Provide facilities for the Employer's Personnel as stated in the Employer's Requirements. c) Not to permit any of the Contractor's Personnel to maintain any temporary or permanent living quarters within the structures forming part of the Permanent Works.	None.	None.
6.7 Health and Safety	a) Take all reasonable precautions to maintain the health and safety of the Contractor's Personnel. b) Ensure that medical staff, first aid facilities, sick bay and ambulance service are available at all times and that suitable arrangements are made for all necessary welfare and hygiene requirements and for the prevention of epidemics. c) Appoint an accident prevention officer and whatever is required by this person to exercise this responsibility and authority. d) Send to the Engineer, details of any accident. e) Maintain records and make reports concerning health, safety, welfare and damage to property.	a) None. b) None. c) None. d) As soon as practicable after its occurrence. e) None.	None.
6.8 Contractor's Superintendence	Provide all necessary superintendence to plan, arrange, direct, manage, inspect and test the work.	None.	None.
6.10 Records of Contractor's Personnel and Equipment	Submit to the Engineer, details showing the number of each class of Contractor's Personnel and of each type of Contractor's Equipment on the Site.	Each calendar month.	None.

CLAUSE	OBLIGATIONS	TIME FRAME	SPECIFIC CONSEQUENCES OF NON-COMPLIANCE
6.11 Disorderly Conduct	Take all reasonable precautions to prevent any unlawful, riotous or disorderly conduct by, or amongst the Contractor's Personnel.	None.	None.
7 Plant, Materials and Workmanship			
7.1 Manner of Execution	Carry out the manufacture of Plant, the production and manufacture of Materials and all other execution of the Works.	None.	None.
7.2 Samples	Submit samples of Materials and relevant information to the Engineer for consent.	None.	None.
7.3 Inspection	a) Give the Employer's Personnel full opportunity to carry out inspections. b) Give notice to the Engineer to inspect.	a) None. b) Whenever any work is ready and before it is covered up, put out of sight or packaged for storage or transport.	a) None. b) Uncover the work, reinstate and make good at the Contractor's cost.
7.4 Testing	a) Provide everything necessary to carry out the specified tests. b) Agree with the Engineer, the time and place for the testing. c) Give notice if the Contractor suffers delay and/or incurs Cost as a result of complying with instructions or a delay for which the Employer is responsible. d) Forward to the Engineer certified reports of the tests.	a) None. b) None. c) As soon as practicable and not later than 28 days after the Contractor became aware, or should have become aware of the event or circumstance (Sub-Clause 20.1). d) Promptly.	a) None. b) None. c) Loss of entitlement to an extension to the Time for Completion and additional payment (Sub-Clause 20.1). d) None.
7.5 Rejection	Make good defects notified by the Engineer.	Promptly.	None.

THE OBLIGATIONS OF THE CONTRACTOR (continued)

CLAUSE	OBLIGATIONS	TIME FRAME	SPECIFIC CONSEQUENCES OF NON-COMPLIANCE
7.6 Remedial Work	Comply with the instructions of the Engineer with regard to remedial work.	Within a reasonable time as specified in the instruction or immediately if urgency is specified.	Contractor shall pay costs incurred by the Employer in engaging other persons to carry out the work.
7.8 Royalties	Pay all royalties, rents and other payments for natural Materials obtained from outside the Site and disposal of surplus materials.	None.	None.
8 Commencement, Delays and Suspension			
8.1 Commencement of Works	Commence the execution of the Works and proceed with the Works with due expedition and without delay.	As soon as is reasonably practicable after the Commencement Date.	None.
8.2 Time for Completion	Complete the whole of the Works and each Section within the times specified in the Contract.	None.	Contractor shall pay delay damages to the Employer (Sub-Clause 8.7).
8.3 Programme	a) Submit a detailed time programme. b) Submit a revised programme. c) Proceed in accordance with the programme. d) Give notice to the Engineer of specific probable future events or circumstances which may adversely affect the work, increase the Contract Price or delay the execution of the Works. e) Submit a revised programme on receiving a notice from the Engineer that a programme fails to comply with the Contract or to be consistent with actual progress.	a) Within 28 days after receiving the notice of commencement. b) Whenever the previous programme is inconsistent with actual progress or with the Contractor's obligations. c) None. d) Promptly. e) None.	None.

CLAUSE	OBLIGATIONS	TIME FRAME	SPECIFIC CONSEQUENCES OF NON-COMPLIANCE
8.4 Extension of Time for Completion	Give notice to the Engineer if the Contractor considers himself to be entitled to an extension of the Time for Completion.	As soon as practicable and not later than 28 days after the Contractor became aware, or should have become aware of the event or circumstance (Sub-Clause 20.1).	Loss of entitlement to an extension to the Time for Completion (Sub-Clause 20.1).
8.6 Rate of Progress	Adopt revised methods in order to expedite progress and complete within the Time for Completion.	None.	None.
8.7 Delay Damages	Pay delay damages in the case of failure to comply with the Time for Completion.	None.	None.
8.8 Suspension of Work	Protect, store and secure such part or the Works in the case of an instruction to suspend the Works.	None.	None.
8.9 Consequences of Suspension	Give notice to the Engineer if the Contractor suffers delay and/or incurs cost as a result of complying with the Engineer's instructions under sub-Clause 8.8.	As soon as practicable and not later than 28 days after the Contractor became aware, or should have become aware of the event or circumstance (Sub-Clause 20.1).	Loss of entitlement to an extension to the Time for Completion and additional payment (Sub-Clause 20.1).
8.12 Resumption of Work	a) Jointly examine the Works and the Plant and Materials affected by the suspension with the Engineer. b) Make good any deterioration, defect or loss.	None.	None.

Yellow Book

THE OBLIGATIONS OF THE CONTRACTOR (continued)

CLAUSE	OBLIGATIONS	TIME FRAME	SPECIFIC CONSEQUENCES OF NON-COMPLIANCE
9 Tests on Completion			
9.1 Contractor's Obligations	a) Carry out the Tests on Completion. b) Give to the Engineer notice of the date after which the Contractor will be ready to carry out each of the Tests on Completion. c) In the case of trial operation, give the Engineer notice that the Works are ready for any other Tests on Completion. d) Submit a certified report of the results of the Tests to the Engineer.	a) After providing the documents in accordance with Sub-Clauses 5.6 and 5.7. b) Not less than 21 days. c) When to Works are operating under stable conditions. d) As soon as the Works or a Section have passed the Tests on Completion.	None.
9.2 Delayed Tests	Carry out the Tests if the Engineer gives notice of undue delay.	Within 21 days of the Engineer's notice.	The Employer's Personnel may proceed with the tests at the Contractor's cost.
10 Employer's Taking Over			
10.2 Taking Over of Parts of the Works	a) Carry out any outstanding Tests on Completion. b) Give notice of Costs incurred as a result of the Employer taking over and/or using a part of the Works.	a) As soon as practicable. b) As soon as practicable and not later than 28 days after the Contractor became aware, or should have become aware, of the event or circumstance (Sub-Clause 20.1).	a) None. b) Loss of entitlement to additional payment (Sub-Clause 20.1).

CLAUSE	OBLIGATIONS	TIME FRAME	SPECIFIC CONSEQUENCES OF NON-COMPLIANCE
10.3 Interference with Tests on Completion	a) In the case of prevention from carrying out the tests, carry out any outstanding Tests on Completion. b) Give notice if the Contractor suffers delay and/or incurs Cost as a result of interference with Tests on Completion.	a) As soon as practicable. b) As soon as practicable and not later than 28 days after the Contractor became aware, or should have become aware of the event or circumstance (Sub-Clause 20.1).	c) None. d) Loss of entitlement to an extension to the Time for Completion and additional payment (Sub-Clause 20.1).
11 Defects Liability			
11.1 Completion of Outstanding Work and Remedying Defects	a) Complete any work which is outstanding on the date stated in a Taking-Over Certificate. b) Execute all work required to remedy defects or damage.	a) Within such reasonable time as is instructed by the Engineer. b) On or before the expiry date of the Defects Notification Period.	a) The Employer may carry out the work himself, at the Contractor's cost (Sub-Clause 11.4). b) A reduction in the Contract Price may be made (Sub-Clause 11.4).
11.8 Contractor to Search	If required by the Engineer, search for the cause of any defect.	None.	None.
11.11 Clearance of Site	Remove any remaining Contractor's Equipment, surplus material, wreckage, rubbish and Temporary Works from the Site.	Within 28 days of receipt of the Performance Certificate.	The Employer may sell or otherwise dispose of any remaining items and the Employer is entitled to recover the costs of disposal.

Yellow Book

THE OBLIGATIONS OF THE CONTRACTOR (continued)

CLAUSE	OBLIGATIONS	TIME FRAME	SPECIFIC CONSEQUENCES OF NON-COMPLIANCE
12 Tests after Completion			
12.1 Procedure for Tests after Completion	Jointly, with the Employer, compile and evaluate the results of the Tests after Completion.	None.	None.
12.2 Delayed Tests	In the case of unreasonable delay by the Employer, give notice to the Employer.	As soon as practicable and not later than 28 days after the Contractor became aware, or should have become aware of the event or circumstance (Sub-Clause 20.1).	Loss of entitlement to additional payment (Sub-Clause 20.1).
12.3 Retesting	If the Works or a Section fail to pass the Tests after Completion, execute all work required to remedy defects or damage.	On or before the expiry date of the Defects Notification Period.	A reduction in the Contract Price may be made (Sub-Clause 11.4).
12.4 Failure to Pass Tests after Completion	In the case of unreasonable delay by the Employer in permitting access to the Works or Plant, give notice of Costs incurred as a result of the Employer taking over and/or using a part of the Works.	As soon as practicable and not later than 28 days after the Contractor became aware, or should have become aware of the event or circumstance (Sub-Clause 20.1).	Loss of entitlement to additional payment (Sub-Clause 20.1).
13 Variations and Adjustments			
13.1 Right to Vary	Execute and be bound by each Variation.	None.	None.
13.3 Variation Procedure	a) Respond in writing to a request for a proposal. b) Not to delay any work whilst awaiting a response. c) Acknowledge receipt of Variation instructions.	a) As soon as practicable. b) None. c) None.	None.

CLAUSE	OBLIGATIONS	TIME FRAME	SPECIFIC CONSEQUENCES OF NON-COMPLIANCE
13.5 Provisional Sums	Produce quotations, invoices, vouchers and accounts or receipts in substantiation of the amounts paid to nominated Subcontractors.	When required by the Engineer.	None.
13.6 Daywork	a) Submit quotations to the Engineer. b) Submit invoices, vouchers and accounts or receipts for Goods. c) Deliver to the Engineer, statements which include the details of the resources used in executing the previous day's work. d) Submit priced statements of these resources.	a) Before ordering Goods for the work to be executed on a Daywork basis. b) When applying for payment. c) Each day. d) Prior to their inclusion in the next Statement under Sub-Clause 14.3.	None.
13.7 Adjustments for Changes in Legislation	Give notice if the Contractor suffers delay and/or incurs Cost as a result of changes in legislation.	As soon as practicable and not later than 28 days after the Contractor became aware, or should have become aware of the event or circumstance (Sub-Clause 20.1).	Loss of entitlement to an extension to the Time for Completion and additional payment (Sub-Clause 20.1).
14 Contract Price and Payment			
14. 1 The Contract Price	Pay all taxes, duties and fees to be paid under the Contract.	None.	None.
14.2 Advance Payment	a) Submit an advance payment guarantee. b) Extend the validity of the guarantee until the advance payment has been repaid.	None.	a) Employer is not obligated to make the advance payment. b) None.
14.3 Application for Interim Payment Certificates	Submit a Statement showing in detail the amounts to which the Contractor considers himself to be entitled.	After the period of payment stated in the Contract, or after the end of each month.	No obligation on the Engineer to certify payment (Sub-Clause 14.6).

Yellow Book

THE OBLIGATIONS OF THE CONTRACTOR (continued)

CLAUSE	OBLIGATIONS	TIME FRAME	SPECIFIC CONSEQUENCES OF NON-COMPLIANCE
14.4 Schedule of Payments	In the case that the Contract does not include a schedule of payments, submit non-binding estimates of the payments expected to become due.	a) First estimate within 42 days after the Commencement Date. b) Revised estimates at quarterly intervals.	None.
14.10 Statement at Completion	Submit a Statement at completion.	Within 84 days after receiving the Taking-Over Certificate for the Works.	None.
14.11 Application for Final Payment Certificate	a) Submit a draft final statement. b) Submit such further information as the Engineer may reasonably require. c) Prepare and submit the Final Statement as agreed with the Engineer.	a) Within 56 days after receiving the Performance Certificate. b) None. c) None.	a) None. b) None. c) The Engineer will certify an amount as he fairly determines to be due.
14.12 Discharge	Submit a written discharge.	When submitting the Final Statement.	None.
15 Termination by Employer			
15.2 Termination by Employer	a) In the case of a notice of termination being served, leave the Site and deliver any required Goods, Contractor's Documents and other design documents, to the Engineer. b) Use best efforts to comply with any reasonable instructions included in the notice. c) Arrange for the removal of Equipment and Temporary Works.	a) None. b) Immediately. c) Promptly.	a) None. b) None. c) Items may be sold by the Employer.
15.5 Employer's Entitlement to Termination	In the case of a notice of termination, cease all further work, hand over Contractor's Documents, Plant, Materials and other work and remove all other Goods from the Site (Sub-Clause 16.3).	28 days from the Employer's notice or return of the Performance Security, whichever is the later.	None.

CLAUSE	OBLIGATIONS	TIME FRAME	SPECIFIC CONSEQUENCES OF NON-COMPLIANCE
16 Suspension and Termination by Contractor			
16.1 Contractor's Entitlement to Suspend Work	a) Give notice if the Contractor intends to suspend work or reduce the rate of work. b) Resume normal working when the Employer's obligations have been met. c) Give further notice if the Contractor suffers delay and/or incurs Cost as a result of suspending work or reducing the rate of work.	a) 21 days before the intended suspension or reduction in the rate of work. b) As soon as is reasonably practicable. c) As soon as practicable and not later than 28 days after the Contractor became aware, or should have become aware of the event or circumstance (Sub-Clause 20.1).	a) None. b) None. c) Loss of entitlement to an extension to the Time for Completion and additional payment (Sub-Clause 20.1).
16.2 Termination by Contractor	Give notice of intention to terminate.	14 days before the intended termination date.	None.
16.3 Cessation of Work and Removal of Contractor's Equipment	In the case of a notice of termination, cease all further work, hand over Contractor's Documents, Plant, Materials and other work and remove all other Goods from the Site.	After the notice has taken effect.	None.
17 Risk and Responsibility			
17.1 Indemnities	Indemnify and hold harmless the Employer, the Employer's Personnel and their respective agents against and from all claims, damages, losses and expenses in respect of bodily injury, sickness, disease, death, damage to or of loss of property by reason of the Contractor's design, the execution and completion of the Works.	None.	None.

THE OBLIGATIONS OF THE CONTRACTOR

Yellow Book

THE OBLIGATIONS OF THE CONTRACTOR (continued)

CLAUSE	OBLIGATIONS	TIME FRAME	SPECIFIC CONSEQUENCES OF NON-COMPLIANCE
17.2 Contractor's Care of the Works	a) Take full responsibility for the care of the Works and Goods. b) Take responsibility for the care of any work which is outstanding on the date stated in a Taking-Over Certificate. c) Rectify loss or damage if any loss or damage happens to the Works, Goods or Contractor's Documents.	a) From the Commencement Date until the Taking-Over Certificate is issued. b) Until the outstanding work has been completed. c) None.	None.
17.4 Consequences of Employer's Risks	a) Give notice in the case of an Employer's risk event which results in loss or damage. b) Rectify the loss or damage as required by the Engineer. c) Give further notice if the Contractor suffers delay and/or incurs Cost as a result of rectifying loss or damage caused by Employers Risks.	a) Promptly. b) None. c) As soon as practicable and not later than 28 days after the Contractor became aware, or should have become aware of the event or circumstance (Sub-Clause 20.1).	a) None. b) None. c) Loss of entitlement to an extension to the Time for Completion and additional payment (Sub-Clause 20.1).
17.5 Intellectual and Industrial Property Rights	a) Indemnify and hold the Employer harmless against and from any other claim which arises out of, or in relation to the Contractor's design, manufacture, construction or execution of the Works, the use of Contractor's equipment or the proper use of the Works. b) If requested by the Employer, assist in contesting the claim. c) Not to make any admission which might be prejudicial to the Employer.	a) None. b) None. c) None.	a) None. b) None. c) None.

CLAUSE	OBLIGATIONS	TIME FRAME	SPECIFIC CONSEQUENCES OF NON-COMPLIANCE
18 Insurance			
18.1 General Requirements for Insurances	a) Wherever the Contractor is the insuring Party, effect and maintain the insurances in terms consistent with any terms agreed by the Parties before the date of the Letter of Acceptance. b) Act under the policy on behalf of any additional joint insured parties. c) Submit evidence to the Employer that the insurances have been effected and copies of the policies. d) Submit evidence of payment of premiums. e) Inform the insurers of any relevant changes to the execution of the Works and ensure that insurance is maintained. f) Not to make any material alteration to the terms of any insurance without approval of the Employer.	a) Within the periods stated in the Contract Data. b) None. c) Within the periods stated in the Appendix to Tender. d) Upon payment of premium. e) As appropriate. f) None.	Employer may effect the insurances and recover the Cost from the Contractor.
18.4 Insurance for Contractor's Personnel	Effect and maintain insurance against injury, sickness, disease or death of any person employed by the Contractor, or any other of the Contractor's Personnel.	From the time that personnel are assisting in the execution of the Works.	Employer may effect the insurances and recover the Cost from the Contractor (Sub-Clause 18.1).
19 Force Majeure			
19.2 Notice of Force Majeure	Give notice in the case that the Contractor is, or will be prevented from performing any of its obligations under the Contract by Force Majeure.	Within 14 days after the Contractor became aware, or should have become aware of the relevant event or circumstance constituting Force Majeure.	Contractor shall not be excused performance of the obligations.

THE OBLIGATIONS OF THE CONTRACTOR (continued)

CLAUSE	OBLIGATIONS	TIME FRAME	SPECIFIC CONSEQUENCES OF NON-COMPLIANCE
19.3 Duty to Minimise Delay	a) Use all reasonable endeavours to minimise any delay in the performance of the Contract as a result of Force Majeure. b) Give notice when the effects of the Force Majeure cease.	None.	None.
20 Claims, Disputes and Arbitration			
20.1 Contractor's Claims	a) Give notice if the Contractor considers himself to be entitled to any extension of the Time for Completion and/or any additional payment. b) Submit any other notices which are required by the Contract and supporting particulars of the claim. c) Keep such contemporary records as may be necessary to substantiate any claim and permit the Engineer to inspect all the records. d) Send to the Engineer a fully detailed claim. e) Send further interim claims if the event or circumstance giving rise to the claim has a continuing effect. f) Send a final claim.	a) As soon as practicable and not later than 28 days after the Contractor became aware, or should have become aware of the event or circumstance. b) None. c) None. d) Within 42 days after the Contractor became aware (or should have become aware) of the event or circumstance giving rise to the claim. e) At monthly intervals. f) Within 28 days after the end of the effects resulting from the event or circumstance.	a,b,d) Loss of entitlement to an extension to the Time for Completion and additional payment. c,e,f) The Employer will take account of the extent to which the failure has prevented or prejudiced proper investigation of the claim.

CLAUSE	OBLIGATIONS	TIME FRAME	SPECIFIC CONSEQUENCES OF NON-COMPLIANCE
20.2 Appointment of the Dispute Adjudication Board	a) Jointly appoint the Dispute Adjudication Board (DAB). b) Mutually agree the terms of remuneration for the DAB. c) Not to act alone in the termination of any member of the DAB.	a) Within 28 days after either Party gives notice of intention to refer a dispute to a DAB. b) None. c) None.	a) The appointing entity or official named in the Appendix to Tender shall appoint (Sub-Clause 20.3). b) None. c) None.
20.4 Obtaining Dispute Adjudication Board's Decision	a) Make available to the DAB, additional information, access to the Site and appropriate facilities as the DAB may require. b) Give effect to a DAB decision unless and until it is revised in an amicable settlement or an arbitral award. c) Continue to proceed with the Works in accordance with the Contract.	a) Promptly. b) Promptly. c) None.	a) None. b&c) The matter may be referred to arbitration (Sub-Clause 20.7).
20.5 Amicable Settlement	In the case of a notice of dissatisfaction being issued, attempt to settle the dispute amicably.	Within 56 days of the notice.	Arbitration may be commenced.
GENERAL CONDITIONS OF DISPUTE ADJUDICATION AGREEMENT			
2 General Provisions	Give notice to the DAB Member that the Dispute Adjudication Agreement has taken effect.	Upon all parties signing the Dispute Adjudication Agreement.	None.
5 General Obligations of the Employer and the Contractor	Not to request advice from, or consult with the Member regarding the Contract, otherwise than in the normal course of the DAB's activities.	None.	None.

THE OBLIGATIONS OF THE CONTRACTOR (continued)

CLAUSE	OBLIGATIONS	TIME FRAME	SPECIFIC CONSEQUENCES OF NON-COMPLIANCE
6 Payment	a) Pay the DAB advance payment for 25% of the estimated fees and expenses. b) Pay the DAB invoices. c) Apply to the Employer for reimbursement of one-half of the DAB invoices by way of the Statements.	a) Upon receipt of the invoice. b) Within 28 days of receipt. c) None.	a) DAB member not obliged to engage in activities under the DAB Agreement. b) DAB member not obliged to render its decision. a&b) Employer entitled to pay fees and recover reimbursement of fees, plus financing charges from the Contractor. a&b) DAB Member may suspend services or resign the appointment. c) The Employer is not obliged to reimburse the Contractor.
Annex – Procedural Rules			
1.	a) Jointly with the Employer, furnish to each member of the DAB, one copy of all documents which the DAB may request. b) Copy the Employer on all communications between the DAB and the Contractor.	None.	None.

THE OBLIGATIONS OF THE ENGINEER

CLAUSE	OBLIGATIONS	TIME FRAME	SPECIFIC CONSEQUENCES OF NON-COMPLIANCE
GENERAL CONDITIONS			
1 General Provisions			
1.3 Communications	a) Not to unreasonably withhold approvals, certificates, consents and determinations. b) When a certificate is issued to a Party, send a copy to the other Party.	None.	None.
1.5 Priority of Documents	In the case that an ambiguity or discrepancy is found in the Contract documents, issue any necessary clarification or instruction.	None.	None.
1.9 Errors in the Employer's Requirements	In the case of a notice and claim for delay or Cost being received, respond to the claim and determine the matters.	Respond within 42 days after receiving a claim or any further particulars supporting a previous claim (Sub-Clause 20.1).	None.
2 The Employer			
2.1 Right of Access to the Site	In the case of a notice and claim for delay or Cost being received, respond to the claim and agree or determine the matters.	Respond within 42 days after receiving a claim or any further particulars supporting a previous claim (Sub-Clause 20.1).	None.
2.5 Employer's Claims	a) In the case that the Employer considers himself to be entitled to any payment, give notice and particulars to the Contractor (the Employer may also undertake this action). b) Agree or determine the matters.	a) As soon as practicable after the Employer became aware of the event or circumstances giving rise to the claim. b) None.	None.

THE OBLIGATIONS OF THE ENGINEER (continued)

CLAUSE	OBLIGATIONS	TIME FRAME	SPECIFIC CONSEQUENCES OF NON-COMPLIANCE
3 The Engineer			
3.1 Engineer's Duties and Authority	a) Carry out the duties assigned in the Contract. b) Provide staff that shall include suitably qualified engineers and other professionals who are competent to carry out these duties. c) Obtain the approval of the Employer before exercising any authority specified in the Particular Conditions.	None.	a) None. b) None. c) The Employer shall be deemed to have given approval.
3.2 Delegation by the Engineer	a) In the case of delegation of the Engineer's authority, delegate such authority in writing. b) Not to delegate the authority to determine any matter in accordance with Sub-Clause 3.5 [Determinations]. c) In the case that the Contractor questions any determination or instruction of an assistant and refers the matter, confirm, reverse or vary the determination or instruction.	a) None. b) None. c) Promptly.	None.
3.3 Instructions of the Engineer	Give instructions in writing.	None.	None.

CLAUSE	OBLIGATIONS	TIME FRAME	SPECIFIC CONSEQUENCES OF NON-COMPLIANCE
3.5 Determinations	a) Consult with each Party in an endeavour to reach agreement. b) If agreement is not achieved, make a fair determination in accordance with the Contract, taking due regard of all relevant circumstances. c) Give notice to both Parties of each agreement or determination, with supporting particulars.	None.	None.
4 The Contractor			
4.2 Performance Security	Cooperate with the Contractor to approve the issuing entity and the form of Performance Security.	None.	None.
4.3 Contractor's Representative	Respond to the Contractor's request for consent to the appointment of the Contractor's Representative.	None.	None.
4.4 Subcontractors	Respond to the Contractor's requests for consent for proposed Subcontractors.	None.	None.
4.7 Setting Out	In the case of a notice and claim for delay or Cost being received, respond to the claim and agree or determine the matters.	Respond within 42 days after receiving a claim or any further particulars supporting a previous claim (Sub-Clause 20.1).	None.
4.12 Unforeseeable Physical Conditions	a) In the case of a notice of unforeseen physical conditions, inspect the physical conditions. b) In the case of a notice and claim for delay or Cost being received, respond to the claim and agree or determine the matters.	a) None. b) Respond within 42 days after receiving a claim or any further particulars supporting a previous claim (Sub-Clause 20.1).	None.
4.19 Electricity, Water and Gas	Agree or determine the quantities and amounts due to the Employer for the Contractor's consumption of electricity, water and gas.	None.	None.

Yellow Book

THE OBLIGATIONS OF THE ENGINEER (continued)

CLAUSE	OBLIGATIONS	TIME FRAME	SPECIFIC CONSEQUENCES OF NON-COMPLIANCE
4.20 Employer's Equipment and Free-Issue Material	Agree or determine the amounts due to the Employer for the Contractor's use of Employer's Equipment.	None.	None.
4.23 Contractor's Operations on Site	Cooperate with the Contractor to agree additional working areas outside the Site.	None.	None.
4.24 Fossils	a) Give instructions for dealing with fossils, coins, articles of value or antiquity, and structures and other remains or items of geological or archaeological interest found on the Site. b) In the case of a notice and claim for Cost or delay being received, respond to the claim and agree or determine the matters.	a) None. b) Respond within 42 days after receiving a claim or any further particulars supporting a previous claim (Sub-Clause 20.1).	None.
5 Design			
5.1 General Design Obligations	a) Respond to the Contractor's requests for consent to proposed design Subcontractors. b) In the case of a notice of any error, fault or other defect found in the Employer's Requirements, determine whether Clause 13 (Variations and Adjustments) shall be applied and give notice to the Contractor.	None.	None.

CLAUSE	OBLIGATIONS	TIME FRAME	SPECIFIC CONSEQUENCES OF NON-COMPLIANCE
5.2 Contractor's Documents	a) Review and/or approve the Contractor's Documents as described in the Employer's Requirements. b) Give notice to the Contractor of approval or otherwise of the Contractor's Documents.	Within 21 days of receipt of the Contractor's Documents and notice, unless otherwise stated in the Employer's Requirements.	The Engineer shall be deemed to have approved the Contractor's Documents and the Contractor may proceed.
5.4 Technical Standards and Regulations	Respond to the Contractor's proposals for compliance with changed or new applicable standards.	None.	None.
5.6 As-Built Documents	a) Review the Contractor's as-built drawings. b) Cooperate with the Contractor to agree the size, the referencing system and other relevant details of the as-built drawings and provide consent.	None.	None.
6 Staff and Labour			
6.10 Records of Contractor's Personnel and Equipment	Cooperate with the Contractor to approve a form to record the number of each class of Contractor's Personnel and of each type of Contractor's Equipment on the Site.	None.	None.
7 Plant, Materials and Workmanship			
7.3 Inspection	Examine, inspect, measure and test the materials and workmanship.	Without unreasonable delay (or promptly give notice that inspection is not required).	None.

THE OBLIGATIONS OF THE ENGINEER (continued)

CLAUSE	OBLIGATIONS	TIME FRAME	SPECIFIC CONSEQUENCES OF NON-COMPLIANCE
7.4 Testing	a) Agree with the Contractor the time and place for specified testing. b) Give the Contractor notice of intention to attend the tests. c) In the case of a notice and claim for delay or cost being received, respond to the claim and agree or determine the matters. d) Endorse the Contractor's test certificate, or issue a certificate confirming tests have been passed.	a) None. b) Not less than 24 hours. c) Respond within 42 days after receiving a claim or any further particulars supporting a previous claim (Sub-Clause 20.1). d) Promptly.	a) None. b) If the Engineer does not attend, the Contractor may proceed and the tests shall be deemed to have been made in Engineer's presence. c) None. d) None.
7.5 Rejection	In the case of Plant, Materials, design or workmanship being found to be defective or not in accordance with the Contract, give notice of rejection.	None.	None.
8 Commencement, Delays and Suspension			
8.1 Commencement of Works	Give the Contractor notice of the Commencement Date.	Not less than 7 days before the Commencement Date and within 42 days after the Contractor receives the Letter of Acceptance.	None.
8.3 Programme	In the case that a programme does not comply with the Contract, give notice to the Contractor.	Within 21 days after receiving the programme.	Contractor shall proceed in accordance with the programme.

CLAUSE	OBLIGATIONS	TIME FRAME	SPECIFIC CONSEQUENCES OF NON-COMPLIANCE
8.4 Extension of Time for Completion	In the case of a notice and claim for delay being received, respond to the claim and agree or determine the matters.	Respond within 42 days after receiving a claim or any further particulars supporting a previous claim (Sub-Clause 20.1).	None.
8.9 Consequences of Suspension	In the case of a notice and claim for delay or Cost being received, respond to the claim and agree or determine the matters.	Respond within 42 days after receiving a claim or any further particulars supporting a previous claim (Sub-Clause 20.1).	None.
8.11 Prolonged Suspension	In the case that the Contractor requests permission to proceed after 84 days of suspension, respond to Contractor's request.	Within 28 days of the request.	Contractor may treat the suspension as an omission or give notice of termination.
8.12 Resumption of Work	Jointly with the Contractor examine the Works and the Plant and Materials affected by the suspension.	After permission or instruction to proceed is given.	None.
9 Tests on Completion			
9.1 Contractor's Obligations	Make allowances for the effect of any use of the Works by the Employer on the performance or other characteristics of the Works.	None.	None.
10 Employer's Taking Over			
10.1 Taking Over of the Works and Sections	Issue the Taking-Over Certificate to the Contractor, or reject the Contractor's application, giving reasons and specifying the work required to be done.	Within 28 days after receiving the Contractor's application.	The Taking-Over Certificate shall be deemed to have been issued.

THE OBLIGATIONS OF THE ENGINEER

Yellow Book

THE OBLIGATIONS OF THE ENGINEER (continued)

CLAUSE	OBLIGATIONS	TIME FRAME	SPECIFIC CONSEQUENCES OF NON-COMPLIANCE
10.2 Taking Over of Parts of the Works	a) In the case that the Employer uses part of the Works and if requested by the Contractor, issue a Taking-Over Certificate for this part. b) In the case of a notice and claim for incurred Cost being received, respond to the claim and agree or determine the matters. c) Determine any reduction in delay damages as a result of a Taking-Over Certificate being issued for a part of the Works.	a) None. b) Respond within 42 days after receiving a claim or any further particulars supporting a previous claim (Sub-Clause 20.1). c) None.	None.
10.3 Interference with Tests on Completion	a) In the case of the Contractor being prevented from carrying out Tests on Completion by the Employer, issue a Taking-Over Certificate accordingly. b) In the case of a notice and claim for Cost or delay being received, respond to the claim and agree or determine the matters.	a) None. b) Respond within 42 days after receiving a claim or any further particulars supporting a previous claim (Sub-Clause 20.1).	None.
11 Defects Liability			
11.4 Failure to Remedy Defects	In the case that the Contractor fails to remedy any defect and if required by the Employer, agree or determine a reasonable reduction in the Contract Price.	None.	None.
11.8 Contractor to Search	In the case that the Contractor has searched for a defect that is found not to be the responsibility of the Contractor, agree or determine the cost of the search.	None.	None.

CLAUSE	OBLIGATIONS	TIME FRAME	SPECIFIC CONSEQUENCES OF NON-COMPLIANCE
11.9 Performance Certificate	Issue the Performance Certificate.	Within 28 days after the latest of the expiry dates of the Defects Notification Periods or as soon thereafter as the Contractor has completed his obligations.	None.
12 Tests after Completion			
12.2 Delayed Tests	In the case of a notice and claim for incurred Cost being received, respond to the claim and agree or determine the matters.	Respond within 42 days after receiving a claim or any further particulars supporting a previous claim (Sub-Clause 20.1).	None.
12.4 Failure to Pass Tests after Completion	In the case of a notice and claim for additional Cost being received, respond to the claim and agree or determine the matters.	Respond within 42 days after receiving a claim or any further particulars supporting a previous claim (Sub-Clause 20.1).	None.
13 Variations and Adjustments			
13.1 Right to Vary	In the case that the Contractor gives notice that the Contractor cannot readily obtain the Goods required for a Variation, cancel, confirm or vary the instruction.	None.	None.
13.3 Variation Procedure	a) Respond to the Contractor's Variation or Value Engineering proposals with approval, disapproval or comments. b) Issue instructions to execute Variations. c) Agree or determine adjustments to the Contract Price and Schedule of Payments for Variations.	a) As soon as practicable after receiving the proposal. b) None. c) None.	a) None. b) The Contractor shall not make any alteration and/or modification of the Permanent Works (Sub-Clause 13.1). c) None.

THE OBLIGATIONS OF THE ENGINEER (continued)

CLAUSE	OBLIGATIONS	TIME FRAME	SPECIFIC CONSEQUENCES OF NON-COMPLIANCE
13.5 Provisional Sums	Give instructions for the use of Provisional Sums.	None.	None.
13.6 Daywork	Sign the Contractor's Daywork statements.	If correct, or when agreed.	None.
13.7 Adjustments for Changes in Legislation	In the case of a notice and claim for delay being received, respond to the claim and agree or determine the matters.	Respond within 42 days after receiving a claim or any further particulars supporting a previous claim (Sub-Clause 20.1).	None.
13.8 Adjustments for Changes in Cost	a) In the case that the cost indices or reference prices stated in the table of adjustment data is in doubt, make a determination. b) In the case that each current cost index is not available, determine a provisional index for the issue of Interim Payment Certificates.	a) None. b) Such that the index may be used for calculations for inclusion in the Payment Certificates.	None.
14 Contract Price and Payment			
14.2 Advance Payment	Issue an Interim Payment Certificate for the first instalment of the advance payment.	After receiving a Statement under Sub-Clause 14.3 and after the Employer receives the Performance Security and an advance payment guarantee.	None.
14.3 Application for Interim Payment Certificates	Cooperate with the Contractor to agree and approve the form of the Statements.	None.	None.

CLAUSE	OBLIGATIONS	TIME FRAME	SPECIFIC CONSEQUENCES OF NON-COMPLIANCE
14.5 Plant and Materials intended for the Works	Determine and certify an amount for Plant and Materials which have been sent to the Site for incorporation in the Permanent Works.	For inclusion in each Interim Payment Certificate.	None.
14.6 Issue of Interim Payment Certificates	a) Issue to the Employer an Interim Payment Certificate. b) In the case that the certified amount would be less than the minimum amount of Interim Payment Certificates stated in the Appendix to Tender, give notice to the Contractor.	a) Within 28 days after receiving a Statement from the Contractor. b) None	If late certification results in the Employer not making payment within the stated period, the Contractor is entitled to receive financing charges (Sub-Clause 14.7).
14.9 Payment of Retention Money	a) Certify the first half of the Retention Money. b) Certify the outstanding balance of the Retention Money.	a) When the Taking-Over Certificate has been issued for the Works. b) Promptly after the latest of the expiry dates of the Defects Notification Periods.	If late certification results in the Employer not making payment within the stated period, the Contractor is entitled to receive financing charges (Sub-Clause 14.7).
14.10 Statement at Completion	Issue to the Employer an Interim Payment Certificate.	Within 28 days after receiving a Statement at Completion.	If late certification results in the Employer not making payment within the stated period, the Contractor is entitled to receive financing charges (Sub-Clause 14.7).

THE OBLIGATIONS OF THE ENGINEER

Yellow Book

THE OBLIGATIONS OF THE ENGINEER (continued)

CLAUSE	OBLIGATIONS	TIME FRAME	SPECIFIC CONSEQUENCES OF NON-COMPLIANCE
14.11 Application for Final Payment Certificate	a) Cooperate with the Contractor to agree and approve the form for the Final Statement. b) In the case that a dispute exists, deliver to the Employer (with a copy to the Contractor) an Interim Payment Certificate for the agreed parts of the draft final statement.	None.	None.
14.13 Issue of Final Payment Certificate	a) Issue, to the Employer the Final Payment Certificate. b) In the case that the Contractor has not applied for a Final Payment Certificate, request the Contractor to do so. c) In the case that the Contractor fails to submit an application within a period of 28 days, issue the Final Payment Certificate for such amount as the Engineer fairly determines to be due.	a) Within 28 days after receiving the Final Statement and written discharge. b) None. c) None.	If late certification results in the Employer not making payment within the stated period, the Contractor is entitled to receive financing charges (Sub-Clause 14.7).
15 Termination by Employer			
15.3 Valuation at Date of Termination	Agree or determine the value of the Works, Goods, Contractor's Documents and any other sums due to the Contractor for work executed in accordance with the Contract.	As soon as practicable after a notice of termination.	None.

CLAUSE	OBLIGATIONS	TIME FRAME	SPECIFIC CONSEQUENCES OF NON-COMPLIANCE
16 Suspension and Termination by Contractor			
16.1 Contractor's Entitlement to Suspend Work	In the case of a notice and claim for delay or Cost being received, respond to the claim and agree or determine the matters.	Respond within 42 days after receiving a claim or any further particulars supporting a previous claim (Sub-Clause 20.1).	None.
17 Risk and Responsibility			
17.4 Consequences of Employer's Risks	In the case of a notice and claim for delay or Cost being received, respond to the claim and agree or determine the matters.	Respond within 42 days after receiving a claim or any further particulars supporting a previous claim (Sub-Clause 20.1).	None.
19 Force Majeure			
19.4 Consequences of Force Majeure	In the case of a notice and claim for delay or Cost being received, respond to the claim and agree or determine the matters.	Respond within 42 days after receiving a claim or any further particulars supporting a previous claim (Sub-Clause 20.1).	None.
19.6 Optional Termination, Payment and Release	In the case of termination, determine the value of the work done and issue a Payment Certificate.	Upon termination.	None.
20 Claims, Disputes and Arbitration			
20.1 Contractor's Claims	a) In the case of a claim being received, respond with approval, or with disapproval and detailed comments. b) Agree or determine the extension of the Time for Completion and/or the additional payment.	a) Within 42 days after receiving the claim or any further particulars supporting a previous claim. b) None.	None.

THE OBLIGATIONS OF THE ENGINEER

Yellow Book

THE OBLIGATIONS OF THE DISPUTE ADJUDICATION BOARD

CLAUSE	OBLIGATIONS	TIME FRAME	SPECIFIC CONSEQUENCES OF NON-COMPLIANCE
GENERAL CONDITIONS			
20 Claims, Disputes and Arbitration			
20.4 Obtaining Dispute Adjudication Board's Decision	Give a reasoned decision on any dispute referred to the DAB.	Within 84 days after receiving a dispute reference or the advance payment, whichever is the later.	Either Party may give a notice of dissatisfaction and commence arbitration.
GENERAL CONDITIONS OF DISPUTE ADJUDICATION AGREEMENT			
3 Warranties	a) Be impartial and independent of the Employer, the Contractor and the Engineer. b) Disclose to the Parties and to the Other Members, any fact or circumstance which might appear inconsistent with his/her warranty and agreement of impartiality and independence.	a) None. b) Promptly.	None.

CLAUSE	OBLIGATIONS	TIME FRAME	SPECIFIC CONSEQUENCES OF NON-COMPLIANCE
4 General Obligations of the Member	a) Have no interest, financial or otherwise in the Parties or the Engineer, nor any financial interest in the Contract. b) Not previously to have been employed as a consultant or otherwise by the Parties or the Engineer, except as disclosed in writing. c) Disclose in writing to the Parties and the Other Members, any professional or personal relationships with any director, officer or employee of the Parties or the Engineer and any previous involvement in the overall project of which the Contract forms part. d) Not, for the duration of the Dispute Adjudication Agreement, be employed as a consultant or otherwise by the Parties or the Engineer, except as may be agreed in writing. e) Comply with the procedural rules and with Sub-Clause 20.4 of the Conditions of Contract. f) Not give advice to the Parties, the Employer's Personnel or the Contractor's Personnel concerning the conduct of the Contract, other than in accordance with the procedural rules. g) Not enter into discussions or make any agreement with the Employer, the Contractor or the Engineer regarding employment by any of them after ceasing to act under the Dispute Adjudication Agreement. h) Ensure his/her availability for all site visits and hearings as are necessary. i) Treat the details of the Contract and all the DAB's activities and hearings as private and confidential.	None.	None.

Yellow Book

THE OBLIGATIONS OF THE DISPUTE ADJUDICATION BOARD (continued)

CLAUSE	OBLIGATIONS	TIME FRAME	SPECIFIC CONSEQUENCES OF NON-COMPLIANCE
6 Payment	a) Submit an invoice for an advance of 25% of the estimated total amount of daily fees and the estimated total of expenses that will be incurred. b) Submit invoices for the balance of daily fees and expenses.	a) Immediately after the DAB Agreement takes effect and before engaging in any activities. b) None.	None.
Annex – Procedural Rules			
1.	Copy all communications between the DAB and the Employer or the Contractor to the other Party.	None.	None.
2(a).	a) Act fairly and impartially as between the Parties. b) Give each of the Parties a reasonable opportunity of putting his case and responding to the other's case.	None.	None.
2(b).	Adopt procedures suitable to the dispute, avoiding unnecessary delay or expense.	None.	None.
3.	In the case of a hearing on the dispute, decide on the date and place for the hearing.	None.	None.
6.	a) Not express any opinions during any hearing concerning the merits of any arguments advanced by the Parties. b) Make and give a decision in accordance with Sub-Clause 20.4, or as otherwise agreed by the Employer and the Contractor in writing. c) If the DAB comprises three persons: I. Convene in private after a hearing. II. Endeavour to reach a unanimous decision.	None.	None.

Chapter 5

The Silver Book
Conditions of Contract for EPC/Turnkey Projects, First Edition 1999

The FIDIC Contracts: Obligations of the Parties, First Edition. Andy Hewitt.
© 2014 John Wiley & Sons, Ltd. Published 2014 by John Wiley & Sons, Ltd.

THE OBLIGATIONS OF THE EMPLOYER

CLAUSE	OBLIGATIONS	TIME FRAME	SPECIFIC CONSEQUENCES OF NON-COMPLIANCE
GENERAL CONDITIONS			
1 General Provisions			
1.8 Care and Supply of Documents	In the case of an error in a document, give notice to the Contractor.	Promptly.	None.
1.9 Confidentiality	Treat the details of the Contract as private and confidential.	None.	None.
1.13 Compliance with Laws	Obtain the planning, zoning or similar permission for the Permanent Works, and any other permissions described in the Employers' Requirements as having been (or being) obtained by the Employer.	None.	None.
2 The Employer			
2.1 Right of Access to the Site	a) Give the Contractor right of access to, and possession of, all parts of the Site. b) Give the Contractor possession of any foundation, structure, plant or means of access if required. c) In the case of a notice and claim for delay or cost being received, agree or determine the matters.	a) Within the time (or times) stated in the Particular Conditions. b) In the time and manner stated in the Employer's Requirements. c) Respond within 42 days after receiving a claim or any further particulars supporting a previous claim (Sub-Clause 20.1).	a&b) Contractor shall be entitled an extension of time and payment of Cost plus reasonable profit. c) None.

THE OBLIGATIONS OF THE EMPLOYER (continued)

CLAUSE	OBLIGATIONS	TIME FRAME	SPECIFIC CONSEQUENCES OF NON-COMPLIANCE
2.2 Permits, Licenses or Approvals	Provide reasonable assistance to the Contractor at the request of the Contractor: a) By obtaining copies of the Laws of the Country which are relevant but not readily available and b) For the Contractor's applications for any permits, licences or approvals required by the Laws of the Country.	None.	None.
2.3 Employer's Personnel	Ensure that the Employer's personnel and the Employer's other contractors, cooperate with the Contractor and take actions similar to those which the Contractor is required to take under Sub-Clause 4.8 [Safety Procedures] and under Sub-Clause 4.18 [Protection of the Environment].	None.	None.
2.4 Employer's Financial Arrangements	Submit reasonable evidence that financial arrangements have been made and are being maintained which will enable the Employer to pay the Contract Price.	Within 28 days after receiving any request from the Contractor.	Contractor entitled to suspend work, reduce the rate of work and to an extension of time and additional payment as a result of such actions (Sub-Clause 16.1).
2.5 Employer's Claims	a) Give notice and particulars to the Contractor, if the Employer considers himself to be entitled to any payment under any Clause of the Conditions or otherwise, in connection with the Contract and/or to any extension of the Defects Notification Period. b) Agree or determine the amount of payment and/or the extension of the Defects Notification Period.	a) As soon as practicable after becoming aware of the event or circumstances giving rise to the claim. b) None.	None.

CLAUSE	OBLIGATIONS	TIME FRAME	SPECIFIC CONSEQUENCES OF NON-COMPLIANCE
3 The Employer's Administration			
3.1 The Employer's Representative	a) In the case that an Employer's Representative is appointed, give notice of the name, address, duties and authority of the Employer's Representative to the Contractor. b) Give the Contractor notice in the case of replacement of the Employer's Representative.	a) None. b) Not less than 14 days.	None.
3.2 Other Employer's Personnel	Give notice to the Contractor in the case that duties are assigned and authority delegated to assistants or if such duties and authorities are revoked.	None	The assignment, delegation or revocation shall not take effect.
3.3 Delegated Persons	In the case that the Contractor questions any determination or instruction of a delegated person and refers the matter, confirm, reverse or vary the determination or instruction.	None.	None.
3.5 Determinations	a) Consult with the Contractor in an endeavour to reach agreement. b) If agreement is not achieved, make a fair determination in accordance with the Contract, taking due regard of all relevant circumstances. c) Give notice to the Contractor of each agreement or determination, with supporting particulars.	None.	None.

THE OBLIGATIONS OF THE EMPLOYER

Silver Book

THE OBLIGATIONS OF THE EMPLOYER (continued)

CLAUSE	OBLIGATIONS	TIME FRAME	SPECIFIC CONSEQUENCES OF NON-COMPLIANCE
4 The Contractor			
4.2 Performance Security	a) Cooperate with the Contractor to agree the entity, country (or other jurisdiction) for the issue of the Performance Security. b) Cooperate with the Contractor to agree the form of Performance Security if not in the form annexed to the Particular Conditions. c) Not make a claim under the Performance Security, except for amounts to which the Employer is entitled under the Contract (as listed). d) Return the Performance Security to the Contractor.	a) None. b) None. c) None. d) Within 21 days after to Contractor becomes entitled to receive the Performance Certificate.	None.
4.3 Contractor's Representative	Respond to the Contractor's request for consent to the appointment of the Contractor's Representative.	None.	None.
4.10 Site Data	a) Make available to the Contractor all relevant data in the Employer's possession on sub-surface and hydrological conditions at the Site, including environmental aspects. b) Make available to the Contractor all such data which come into the Employer's possession after the Base Date.	a) Prior to the Base Date. b) None.	None.
4.19 Electricity, Water and Gas	a) Give notice and particulars to the Contractor of the quantities consumed and the amounts due for supplies of electricity, water, gas and other services. b) Agree or determine the amount due for payment by the Contractor.	a) As soon as practicable. b) None.	None.

CLAUSE	OBLIGATIONS	TIME FRAME	SPECIFIC CONSEQUENCES OF NON-COMPLIANCE
4.20 Employer's Equipment and Free-Issue Material	a) Make the Employer's Equipment (if any) available for the use of the Contractor in the execution of the Works in accordance with the details, arrangements and prices stated in the Employer's Requirements. b) Give notice and particulars to the Contractor of the quantities and the amounts due for the use of Employer's Equipment. c) Agree or determine the amount due for payment by the Contractor for the use of Employer's Equipment. d) Supply, free of charge, the "free-issue materials" (if any) in accordance with the details stated in the Employer's Requirements. e) Rectify any notified shortage, defect or default in the free-issue materials.	a) None. b) As soon as practicable. c) None. d) As specified in the Contract. e) Immediately.	None.
4.24 Fossils	a) Take care of and exercise authority over fossils, coins, articles of value or antiquity and structures and other remains, or items of geological or archaeological interest found on the Site. b) In the case of a notice and claim for delay or cost being received, agree or determine the matters.	a) None. b) Respond within 42 days after receiving a claim or any further particulars supporting a previous claim (Sub-Clause 20.1).	None.
5 Design			
5.2 Contractor's Documents	a) Review and/or approve the Contractor's Documents as described in the Employer's Requirements.	Within 21 days of receipt of the Contractor's Documents and notice, unless otherwise stated in the Employer's Requirements.	None.
5.6 As-Built Documents	Cooperate with the Contractor to agree the size, the referencing system and other relevant details of the as-built drawings, and provide consent.	None.	None.

Silver Book

THE OBLIGATIONS OF THE EMPLOYER (continued)

CLAUSE	OBLIGATIONS	TIME FRAME	SPECIFIC CONSEQUENCES OF NON-COMPLIANCE
6 Staff and Labour			
6.10 Records of Contractor's Personnel and Equipment	Cooperate with the Contractor to approve a form to record the number of each class of Contractor's Personnel and of each type of Contractor's Equipment on the Site.	None.	None.
7 Plant, Materials and Workmanship			
7.3 Inspection	Examine, inspect, measure and/or test materials and workmanship.	Without unreasonable delay (or promptly give notice that inspection is not required).	None.
7.4 Testing	a) Agree with the Contractor the time and place for the specified testing. b) Give the Contractor notice of intention to attend the tests. c) In the case of a notice and claim for delay or Cost being received, respond to the claim and agree or determine the matters. d) Endorse the Contractor's test certificate, or issue a certificate confirming that the tests have been passed.	a) None. b) Not less than 24 hours. c) Respond within 42 days after receiving a claim or any further particulars supporting a previous claim (Sub-Clause 20.1). d) Promptly.	a) None. b) If the Employer does not attend, the Contractor may proceed and the tests shall be deemed to have been made in the Employer's presence. c) None. d) None.
7.5 Rejection	In the case of Plant, Materials, design or workmanship being found to be defective or not in accordance with the Contract, give notice of rejection.	None.	None.

CLAUSE	OBLIGATIONS	TIME FRAME	SPECIFIC CONSEQUENCES OF NON-COMPLIANCE
8 Commencement, Delays and Suspension			
8.1 Commencement of Works	Give the Contractor notice of the Commencement Date.	Not less than 7 days before the Commencement Date and within 42 days after the Contract comes into full force.	None.
8.4 Extension of Time for Completion	In the case of a notice and claim for delay being received, respond to the claim and agree or determine the matters.	Respond within 42 days after receiving a claim or any further particulars supporting a previous claim (Sub-Clause 20.1).	None.
8.9 Consequences of Suspension	In the case of a notice and claim for delay or Cost being received, respond to the claim and agree or determine the matters.	Respond within 42 days after receiving a claim or any further particulars supporting a previous claim (Sub-Clause 20.1).	None.
8.12 Resumption of Work	Jointly with the Contractor, examine the Works and the Plant and Materials affected by the suspension.	After permission or instruction to proceed is given.	None.
10 Employer's Taking Over			
10.1 Taking Over of the Works and Sections	a) Take over the Works. b) Issue a Taking-Over Certificate or reject the Contractor's application with reasons and specifying the work required to be done to enable the Taking-Over Certificate to be issued.	a) When the Works have been completed in accordance with the Contract and a Taking-Over Certificate has been issued. b) Within 28 days of the Contractor's notice of application for a Taking-Over Certificate.	a) None. b) If the Works are substantially in accordance with the Contract, the Taking-Over Certificate shall be deemed to have been issued.
10.3 Interference with Tests on Completion	In the case of a notice and claim for delay or Cost being received, agree or determine the matters.	Respond within 42 days after receiving a claim or any further particulars supporting a previous claim (Sub-Clause 20.1).	None.

THE OBLIGATIONS OF THE EMPLOYER

Silver Book

THE OBLIGATIONS OF THE EMPLOYER (continued)

CLAUSE	OBLIGATIONS	TIME FRAME	SPECIFIC CONSEQUENCES OF NON-COMPLIANCE
11 Defects Liability			
11.2 Cost of Remedying Defects	Notify the Contractor of any work to be remedied, if due to any cause outside the provisions of the Contract.	None.	None.
11.4 Failure to Remedy Defects	In the case of failure by the Contractor to remedy any defect or damage, notify the Contractor of the date by which the defect or damage is to be remedied.	Within reasonable time.	None.
11.8 Contractor to Search	In the case that the defect is not to be remedied at the cost of the Contractor, agree or determine the cost of the search.	None.	None.
11.9 Performance Certificate	Issue the Performance Certificate.	Within 28 days after the latest of the expiry dates of the Defects Notification Periods or as soon thereafter as the Contractor has completed his obligations.	a) The Performance Certificate shall be deemed to have been issued on the date, 28 days after it should have been issued. b) The Contractor shall not be obliged to clear the site. c) Any Employer's liability to the Contractor shall not cease.

CLAUSE	OBLIGATIONS	TIME FRAME	SPECIFIC CONSEQUENCES OF NON-COMPLIANCE
12 Tests after Completion			
12.1 Procedure for Tests after Completion	a) If Tests after Completion are specified, provide all electricity, fuel and materials and make the Personnel and Plant available. b) Give the Contractor notice of the date when the Tests after Completion will be carried out.	a) None. b) 21 days before the date of the Tests.	None.
12.2 Delayed Tests	In the case of a notice and claim for Cost being received, agree or determine the matters.	Respond within 42 days after receiving a claim or any further particulars supporting a previous claim (Sub-Clause 20.1).	None.
12.4 Failure to Pass Tests after Completion	In the case of a notice and claim for Cost being received, agree or determine the matters.	Respond within 42 days after receiving a claim or any further particulars supporting a previous claim (Sub-Clause 20.1).	None.
13 Variations and Adjustments			
13.1 Right to Vary	In the case that the Contractor gives notice that the Contractor cannot readily obtain the Goods required for a Variation, cancel, confirm or vary the instruction.	None.	None.
13.3 Variation Procedure	a) Respond to the Contractor's Variation or Value Engineering proposals with approval, disapproval or comments. b) Issue instructions to execute Variations. c) Agree or determine adjustments to the Contract Price and Schedule of Payments for Variations.	a) As soon as practicable after receiving the proposal. b) None. c) None.	None.
13.5 Provisional Sums	Give instructions for the use of Provisional Sums.	None.	None.
13.6 Daywork	Sign the Contractor's Daywork statements and return to the Contractor.	If correct, or when agreed.	None.

Silver Book

THE OBLIGATIONS OF THE EMPLOYER (continued)

CLAUSE	OBLIGATIONS	TIME FRAME	SPECIFIC CONSEQUENCES OF NON-COMPLIANCE
13.7 Adjustments for Changes in Legislation	In the case of a notice and claim for delay being received, respond to the claim and agree or determine the matters.	Respond within 42 days after receiving a claim or any further particulars supporting a previous claim (Sub-Clause 20.1).	None.
14 Contract Price and Payment			
14.2 Advance Payment	a) Make the advance payment as an interest-free loan for mobilisation. b) Pay the first instalment. c) Cooperate with the Contractor to approve the form of the advance payment guarantee.	a) When the Contractor submits a guarantee in accordance with this Sub-Clause. b) After receiving a Sub-Clause 14.3 Statement, the Performance Security and an advance payment guarantee. c) None.	None.
14.3 Application for Interim Payments	Cooperate with the Contractor to agree and approve the form of the Statements.	None.	None.
14.5 Plant and Materials intended for the Works	Cooperate with the Contractor to approve the form of a bank guarantee for shipped Plant and Materials.	None.	None.
14.6 Interim Payments	Give notice to the Contractor, with supporting particulars, of any items in the Statement with which the Employer disagrees.	Within 28 days after receiving the Statement.	None.

CLAUSE	OBLIGATIONS	TIME FRAME	SPECIFIC CONSEQUENCES OF NON-COMPLIANCE
14.7 Timing of Payments	a) Pay the first instalment of the advance payment. b) Pay the amount due in respect of each Statement. c) Pay the amount due in respect of the Final Statement.	a) Within 42 days after the Contract came into full force or within 21 days after receiving the documents in accordance with Sub-Clause 4.2 [Performance Security] and Sub-Clause 14.2 [Advance Payment], whichever is later. b) Within 56 days after receipt of the Statement and supporting documents. c) Within 42 days after receipt of the Final Statement and written discharge.	1) Contractor entitled to suspend work, reduce the rate of work and to an extension of time and additional payment as a result of such actions (Sub-Clause 16.1). 2) Payment of financing charges to the Contractor (Sub-Clause 14.8).
14.9 Payment of Retention Money	a) Pay the first half of the Retention Money. b) Pay the outstanding balance of the Retention Money.	a) When the Taking-Over Certificate has been issued for the Works. b) Promptly, after the latest of the expiry dates of the Defects Notification Periods.	The Contractor is entitled to receive financing charges (Sub-Clause 14.7).
14.10 Statement at Completion	a) Give notice to the Contractor, with supporting particulars, of any items in the Statement with which the Employer disagrees. b) Pay the Contractor.	a) Within 28 days after receiving the Statement at completion. b) Within 56 days after receipt of the Statement and supporting documents.	The Contractor is entitled to receive financing charges (Sub-Clause 14.7).
14.11 Application for Final Payment	a) Cooperate with the Contractor to agree and approve the form for the Final Statement. b) In the case of a dispute with regard to the draft final statement, pay the agreed parts of the statement.	a) None. b) Within 42 days after receipt of the Final Statement and written discharge.	The Contractor is entitled to receive financing charges (Sub-Clause 14.7).

THE OBLIGATIONS OF THE EMPLOYER

THE OBLIGATIONS OF THE EMPLOYER (continued)

CLAUSE	OBLIGATIONS	TIME FRAME	SPECIFIC CONSEQUENCES OF NON-COMPLIANCE
14.13 Final Payment	Pay the amount finally due, less all amounts previously paid and any deductions in accordance with Sub-Clause 2.5 (Employer's Claims).	Within 42 days after receipt of the Final Statement and written discharge.	The Contractor is entitled to receive financing charges (Sub-Clause 14.7).
14.15 Currencies of Payment	Pay the Contractor in the currency or currencies named in the Contract Agreement.	None.	None.
15 Termination by Employer			
15.2 Termination by Employer	a) Give notice of intention to terminate the Contract. b) Give notice of release of the Contractor's Equipment and Temporary Works.	a) 14 days prior to termination date, or immediately in the case of the Contractor becoming bankrupt or gives or offers bribes or gratuities (or similar as defined in this clause). b) On completion of the Works.	None.
15.3 Valuation at Date of Termination	Agree or determine the value of the Works, Goods, Contractor's Documents and any other sums due to the Contractor for work executed in accordance with the Contract.	As soon as practicable after a notice of termination.	None.
15.4 Payment after Termination	Pay the balance due to the Contractor after recovering any losses, damages and extra costs.	None.	None.
15.5 Employer's Entitlement to Termination	a) Give notice of intention to terminate the Contract. b) Return the Performance Security. c) Not to terminate the Contract in order to execute the Works himself or to arrange for the Works to be executed by another contractor.	a) 28 days prior to the termination date. b) None. c) None.	None.

CLAUSE	OBLIGATIONS	TIME FRAME	SPECIFIC CONSEQUENCES OF NON-COMPLIANCE
16 Suspension and Termination by Contractor			
16.1 Contractor's Entitlement to Suspend Work	In the case of a notice and claim for delay or Cost being received, agree or determine the matters.	Respond within 42 days after receiving a claim or any further particulars supporting a previous claim (Sub-Clause 20.1).	None.
16.4 Payment on Termination	a) Return the Performance Security to the Contractor. b) Pay the Contractor in accordance with Sub-Clause 19.6 *[Optional Termination, Payment and Release].* c) Pay to the Contractor the amount of any loss of profit or other loss or damage sustained by the Contractor as a result of this termination.	Promptly.	None.
17 Risk and Responsibility			
17.1 Indemnities	Indemnify and hold harmless the Contractor, the Contractor's Personnel and their respective agents against and from all claims, damages, losses and expenses in respect of bodily injury, disease or death which is attributable to any negligence, wilful act or breach of the Contract by the Employer, the Employer's Personnel or agents.	None.	None.
17.4 Consequences of Employer's Risks	In the case of a notice and claim for delay or Cost being received, agree or determine the matters.	Respond within 42 days after receiving a claim or any further particulars supporting a previous claim (Sub-Clause 20.1).	None.

THE OBLIGATIONS OF THE EMPLOYER (continued)

CLAUSE	OBLIGATIONS	TIME FRAME	SPECIFIC CONSEQUENCES OF NON-COMPLIANCE
17.5 Intellectual and Industrial Property Rights	a) Indemnify and hold the Contractor harmless against and from any claim, which is or was an unavoidable result of the Contractor's compliance with the Contract or as a result of any Works being used by the Employer. b) If requested by the Contractor, assist in contesting the claim. c) Not make any admission, which might be prejudicial to the Contractor.	None.	None.
18 Insurance			
18.1 General Requirements for Insurances	a) Cooperate with the Contractor to approve the terms of insurances. b) Effect and maintain the insurances in terms consistent with the details annexed to the Particular Conditions, wherever the Employer is the insuring Party. c) Submit evidence that the insurance has been effected, provide copies of the policies and submit evidence of payment. d) Inform the insurers of any relevant changes to the execution of the Works and ensure that insurance is maintained. e) Not make any material alteration to the terms of any insurance without the approval of the Contractor.	a) None. b) None. c) Within the time frames stipulated in the Particular Conditions. d) As appropriate. e) None.	a) None. b) None. c) The Contractor may effect the insurance and the Contract Price shall be adjusted. d) None. e) None.

CLAUSE	OBLIGATIONS	TIME FRAME	SPECIFIC CONSEQUENCES OF NON-COMPLIANCE
19 Force Majeure			
19.2 Notice of Force Majeure	Give notice to the Contractor in the case that the Employer is, or will be prevented from performing the Employer's obligations by Force Majeure.	Within 14 days of becoming aware of the event.	None.
19.3 Duty to Minimise Delay	a) Use all reasonable endeavours to minimise any delay in the performance of the Contract as a result of Force Majeure. b) Give notice when the effects of the Force Majeure cease.	None.	None.
19.4 Consequences of Force Majeure	In the case of a notice and claim for delay or Cost being received, agree or determine the matters.	Respond within 42 days after receiving a claim or any further particulars supporting a previous claim (Sub-Clause 20.1).	None.
19.6 Optional Termination, Payment and Release	In the case of termination due to Force Majeure, pay the Contractor for work carried out and other Costs listed in this sub-clause.	Upon termination.	None.
20 Claims, Disputes and Arbitration			
20.1 Contractor's Claims	a) In the case of a claim being received, respond with approval, or with disapproval and detailed comments. b) Agree or determine the extension of the Time for Completion and/or the additional payment.	a) Within 42 days after receiving the claim or any further particulars supporting a previous claim (Sub-Clause 20.1). b) None.	None.

THE OBLIGATIONS OF THE EMPLOYER (continued)

CLAUSE	OBLIGATIONS	TIME FRAME	SPECIFIC CONSEQUENCES OF NON-COMPLIANCE
20.2 Appointment of the Dispute Adjudication Board	a) Jointly appoint the Dispute Adjudication Board (DAB). b) Not to consult the DAB without the agreement of the Contractor. c) Not to act alone in the termination of any member of the DAB.	a) Within 28 days after a Party gives notice of its intention to refer a dispute to a DAB. b) None. c) None.	a) The appointing entity or official named in the Appendix to Tender shall appoint the DAB (Sub-Clause 20.3). b) None. c) None.
20.4 Obtaining Dispute Adjudication Board's Decision	a) Make available to the DAB, additional information, access to the Site, and appropriate facilities as the DAB may require. b) Give effect to a DAB decision, unless and until it is revised in an amicable settlement or an arbitral award.	a) Promptly. b) None.	a) None. b) The matter may be referred to arbitration (Sub-Clause 20.7).
20.5 Amicable Settlement	In the case of a notice of dissatisfaction being issued, attempt to settle the dispute amicably.	Within 56 days of the notice.	Arbitration may be commenced.

CLAUSE	OBLIGATIONS	TIME FRAME	SPECIFIC CONSEQUENCES OF NON-COMPLIANCE
GENERAL CONDITIONS OF DISPUTE ADJUDICATION AGREEMENT			
2 General Provisions	Give notice to the DAB Member that the Dispute Adjudication Agreement has taken effect.	When all parties have signed the Dispute Adjudication Agreement.	None.
5 General Obligations of the Employer and the Contractor	Not to request advice from or consultation with a DAB Member regarding the Contract otherwise than in the normal course of the DAB's activities.	None.	None.
6 Payment	a) Pay one half of the DAB fees to the Contractor. b) In the case of the Contractor failing to pay the DAB Member, pay the due fees.	a) Within the monthly payments to the Contractor. b) None.	The DAB Member may suspend services or resign the appointment.
Annex – Procedural Rules			
1.	a) Jointly with the Contractor, furnish to each DAB member one copy of all documents which the DAB may request. b) Copy all communications with the DAB to the Contractor.	None.	None.

THE OBLIGATIONS OF THE CONTRACTOR

CLAUSE	OBLIGATIONS	TIME FRAME	SPECIFIC CONSEQUENCES OF NON-COMPLIANCE
GENERAL CONDITIONS			
1 Definitions			
1.8 Care and Supply of Documents	a) Supply to the Employer six copies of each of the Contractor's Documents. b) Keep, on the Site, a copy of the Contract, publications named in the Employer's Requirements, the Contractor's Documents and Variations and other communications given under the Contract. c) In the case of an error in a document, give notice to the Employer.	a) None. b) None. c) Promptly.	None.
1.9 Confidentiality	Treat the details of the Contract as private and confidential.	None.	None.
1.11 Contractor's Use of Employer's Documents	Not to, without consent, copy, use or communicate the Employer's Requirements and other documents of the Employer to a third party, except as necessary.	None.	None.
1.12 Confidential Details	Disclose all such confidential and other information as the Employer may reasonably require in order to verify the Contractor's compliance with the Contract.	None.	None.

CLAUSE	OBLIGATIONS	TIME FRAME	SPECIFIC CONSEQUENCES OF NON-COMPLIANCE
1.13 Compliance with Laws	a) Comply with applicable Laws. b) Give all notices, pay all taxes, duties and fees, and obtain all permits, licences and approvals as required by the Laws.	None.	None.
1.14 Joint and Several Liability	a) In the case of a joint venture, consortium or other unincorporated grouping of two or more persons, notify the Employer of the leader. b) Not to alter the composition of the joint venture or legal status without the prior consent of the Employer.	None.	None.
2 The Employer			
2.1 Right of Access to the Site	Give notice to the Employer if the Contractor suffers delay and/or incurs Cost as a result of failure to give right of access and possession of the Site.	As soon as practicable and not later than 28 days after the Contractor became aware, or should have become aware of the event or circumstance (Sub-Clause 20.1).	Loss of entitlement to an extension to the Time for Completion and additional payment (Sub-Clause 20.1).
3 The Employer's Administration			
3.4 Instructions	Take instructions from the Employer, the Employer's Representative, or from an assistant to whom the appropriate authority has been delegated.	None.	None.

Silver Book

THE OBLIGATIONS OF THE CONTRACTOR (continued)

CLAUSE	OBLIGATIONS	TIME FRAME	SPECIFIC CONSEQUENCES OF NON-COMPLIANCE
4 The Contractor			
4.1 Contractor's General Obligations	a) Design, execute and complete the Works in accordance with the Contract and remedy any defects in the Works. b) Provide the required Plant and Contractor's Documents specified in the Contract and all Contractor's Personnel, Goods, consumables and other things and services, whether of a temporary or permanent nature. c) Be responsible for the adequacy, stability and safety of all Site operations and of all methods of construction. d) Submit details of the arrangements and methods proposed for the execution of the Works. e) Not make significant alteration to the arrangements and methods without previous notice to the Employer.	a) None. b) None. c) None. d) Whenever required by the Employer. e) None.	None.
4.2 Performance Security	a) Obtain a Performance Security for proper performance and deliver it to the Employer. b) Ensure that the Performance Security is valid and enforceable until the Contractor has executed and completed the Works and remedied any defects. c) Extend the validity of the Performance Security until the Works have been completed and any defects have been remedied.	a) Within 28 days after receiving the Letter of Acceptance. b) None. c) As required.	Employer may claim the full amount of the Performance Security.

CLAUSE	OBLIGATIONS	TIME FRAME	SPECIFIC CONSEQUENCES OF NON-COMPLIANCE
4.3 Contractor's Representative	a) Appoint the Contractor's Representative and give him all authority necessary to act on the Contractor's behalf under the Contract. b) Submit to the Employer for consent, the name and particulars of the person the Contractor proposes to appoint. c) If consent is withheld or subsequently revoked, or if the appointed person fails to act, submit the name and particulars of another suitable person for such appointment. d) Not to, without the prior consent of the Employer, revoke the appointment of the Contractor's Representative or appoint a replacement.	a) None. b) Prior to the Commencement Date. c) None. d) None.	None.
4.4 Subcontractors	a) Not subcontract the whole of the Works. b) Be responsible for the acts or defaults of any Subcontractor, his agents or employees. c) Give the Employer notice of the intended appointment of the Subcontractor with detailed particulars, including his experience. d) The intended date of the commencement of the Subcontractor's work. e) The intended date of the commencement of the Subcontractor's work on the Site.	a) None. b) None. c–e) Not less than 28 days.	None.

THE OBLIGATIONS OF THE CONTRACTOR (continued)

CLAUSE	OBLIGATIONS	TIME FRAME	SPECIFIC CONSEQUENCES OF NON-COMPLIANCE
4.6 Co-operation	a) Allow appropriate opportunities for carrying out work to the Employer's Personnel, any other contractors employed by the Employer and the personnel of any legally constituted public authorities. b) Be responsible for the Contractor's construction activities and co-ordinate activities with those of other contractors to the extent specified in the Employer's Requirements. c) Submit such documents which require the Employer to give to the Contractor possession of any foundation, structure, plant or means of access.	a) None. b) None. c) In the time and manner stated in the Employer's Requirements.	None.
4.7 Setting Out	a) Set out the Works in relation to original points, lines and levels of reference specified in the Contract. b) Be responsible for the correct positioning of all parts of the Works.	None.	None.
4.8 Safety Procedures	a) Comply with all applicable safety regulations. b) Take care for the safety of all persons entitled to be on the Site. c) Use reasonable efforts to keep the Site and Works clear of unnecessary obstruction. d) Provide fencing, lighting, guarding and watching of the Works. e) Provide any Temporary Works which may be necessary for the use and protection of the public and of owners and occupiers of adjacent land.	None.	None.

CLAUSE	OBLIGATIONS	TIME FRAME	SPECIFIC CONSEQUENCES OF NON-COMPLIANCE
4.9 Quality Assurance	Institute a quality assurance system and submit details to the Employer.	Before each design and execution stage is commenced.	None.
4.10 Site Data	Be responsible for verifying and interpreting data provided by the Employer on subsurface and hydrological conditions at the Site, including environmental aspects.	None.	None.
4.13 Rights of Way and Facilities	a) Bear all costs and charges for special and/or temporary rights of way. b) Obtain any additional facilities outside the Site which the Contractor may require for the purposes of the Works.	None.	None.
4.14 Avoidance of Interference	Not to interfere with the convenience of the public, or the access to and use and occupation of all roads and footpaths.	None.	None.
4.15 Access Route	a) Use reasonable efforts to prevent any road or bridge from being damaged. b) Be responsible for any maintenance which may be required for the use of access routes. c) Provide all necessary signs or directions along access routes. d) Obtain any permission which may be required from the relevant authorities for use of routes, signs and directions.	None.	None.
4.16 Transport of Goods	a) Give the Employer notice of the date on which any Plant or a major item of other Goods will be delivered to the Site. b) Be responsible for packing, loading, transporting, receiving, unloading, storing and protecting all Goods and other things required for the Works.	a) 21 days before delivery. b) None.	None.

THE OBLIGATIONS OF THE CONTRACTOR (continued)

CLAUSE	OBLIGATIONS	TIME FRAME	SPECIFIC CONSEQUENCES OF NON-COMPLIANCE
4.17 Contractor's Equipment	Be responsible for all Contractor's Equipment.	None.	None.
4.18 Protection of the Environment	a) Take all reasonable steps to protect the environment and to limit damage and nuisance to people and property. b) Ensure that emissions, surface discharges and effluent shall not exceed the values indicated in the Employer's Requirements and shall not exceed the values prescribed by applicable Laws.	None.	None.
4.19 Electricity, Water and Gas	a) Be responsible for the provision of all power, water and other services. b) Provide any apparatus necessary for use of services as may be available on the Site and for measuring the quantities consumed. c) Pay the Employer for the use of services available on the Site.	None.	None.
4.20 Employer's Equipment and Free-Issue Materials	a) Be responsible for the Employer's Equipment when used by the Contractor. b) Pay the Employer for the use of the Employer's Equipment. c) Inspect free-issue materials. d) Give notice of any shortage, defect or default in the free-issue materials.	a) None. b) None. c) None. d) Promptly.	None.

CLAUSE	OBLIGATIONS	TIME FRAME	SPECIFIC CONSEQUENCES OF NON-COMPLIANCE
4.21 Progress Reports	Prepare and submit monthly progress reports.	Monthly, within 7days of the period to which the report relates.	None.
4.22 Security of the Site	Keep unauthorised persons off the Site.	None.	None.
4.23 Contractor's Operations on Site	a) Confine operations to the Site and to any additional areas agreed as working areas. b) Take all necessary precautions to keep the Contractor's Equipment and Contractor's Personnel within the Site and any agreed working areas. c) Keep the Site free from all unnecessary obstruction. d) Store or dispose of any Contractor's Equipment or surplus materials. e) Clear away and remove from the Site any wreckage, rubbish and Temporary Works. f) Leave the Site and the Works in a clean and safe condition.	None.	None.
4.24 Fossils	a) Take reasonable precautions to prevent the Contractor's Personnel or other persons from removing or damaging fossils, coins, articles of value or antiquity, structures and other remains or items of geological or archaeological interest. b) Give notice of the finding of such items. c) Give further notice if the Contractor suffers delay and/or incurs Cost.	a) None. b) Upon discovery. c) As soon as practicable and not later than 28 days after the Contractor became aware, or should have become aware of the event or circumstance (Sub-Clause 20.1).	a) None. b) None. c) Loss of entitlement to an extension to the Time for Completion and additional payment (Sub-Clause 20.1).

THE OBLIGATIONS OF THE CONTRACTOR (continued)

CLAUSE	OBLIGATIONS	TIME FRAME	SPECIFIC CONSEQUENCES OF NON-COMPLIANCE
5 Design			
5.1 General Design Obligations	Carry out and be responsible for the design of the Works.	None.	None.
5.2 Contractor's Documents	a) Prepare all Contractor's Documents. b) Prepare all other documents necessary to instruct the Contractor's Personnel. c) In the case that the Employer's Requirements require it, submit the Contractor's Documents to the Employer for review. d) Give notice of the submission of Contractor's Documents to the Employer for review. e) In the case that the Employer gives notice that a Contractor's Document does not comply with the Contract, rectify and resubmit the document for review. f) Not to commence execution of such part of the Works until the review period has expired. g) Execute parts of the Works in accordance with the documents submitted for review. h) Give notice to the Employer in the case that the Contractor wishes to modify any design or document which has previously been submitted for review. i) Submit revised documents to the Employer in accordance with the Sub-Clause 5.2 procedure.	a) None. b) None. c) None. d) None. e) None. f) None. g) None. h) Immediately. i) None.	None.

CLAUSE	OBLIGATIONS	TIME FRAME	SPECIFIC CONSEQUENCES OF NON-COMPLIANCE
5.3 Contractor's Undertaking	Design, provide the Contractor's Documents, execute and provide the completed Works in accordance with the Laws and the documents forming the Contract, as altered or modified by Variations.	None.	None.
5.4 Technical Standards and Regulations	a) Design, provide the Contractor's Documents, execute and provide the completed Works in accordance with the Country's technical standards, building, construction and environmental Laws, Laws applicable to the product being produced from the Works and other standards specified in the Employer's Requirements. b) Give notice to the Employer in the case that new or applicable standards come into force after the Base Date and if appropriate, submit proposals for compliance.	None.	None.
5.5 Training	Carry out the training of Employer's Personnel in the operation and maintenance of the Works.	None.	None.
5.6 As-Built Documents	a) Prepare and keep up-to-date a complete set of as-built records of the execution of the Works. b) Provide two copies of the as-built records to the Employer. c) Supply and submit to the Employer for review, as-built drawings of the Works. d) Obtain the consent of the Employer as to the size, referencing system and other relevant details of the as-built drawings. e) Supply to the Employer the specified numbers of as-built drawings.	a) None. b) Prior to the commencement of the Tests on Completion. c) None. d) None. e) Prior to the issue of any Taking-Over Certificate.	a) None. b) None. c) None. d) None. e) The Works shall not be considered to be completed for the purposes of taking-over.

THE OBLIGATIONS OF THE CONTRACTOR (continued)

CLAUSE	OBLIGATIONS	TIME FRAME	SPECIFIC CONSEQUENCES OF NON-COMPLIANCE
5.7 Operation and Maintenance Manuals	a) Supply to the Employer, provisional operation and maintenance manuals. b) Supply to the Employer, final operation and maintenance manuals.	a) Prior to the commencement of the Tests on Completion. b) Prior to taking-over.	a) None. b) The Works shall not be considered to be completed for the purposes of taking-over.
5.8 Design Error	Correct any errors, omissions, ambiguities, inconsistencies, inadequacies or other defects that are found in the Contractor's Documents.	None.	None.
6 Staff and Labour			
6.1 Engagement of Staff and Labour	Make arrangements for the engagement of all staff and labour, local or otherwise and for their payment, housing, feeding and transport.	None.	None.
6.2 Rates of Wages and Conditions of Labour	Pay rates of wages and observe conditions of labour which are not lower than those established for the trade or industry where the work is carried out.	None.	None.
6.3 Persons in the Service of Others	Not recruit, or attempt to recruit, staff and labour from amongst the Employer's personnel.	None.	None.
6.4 Labour Laws	a) Comply with all the relevant labour Laws. b) Require employees to obey all applicable Laws.	None.	None.
6.5 Working Hours	Obtain the consent of the Employer if working outside the normal working hours.	None.	None.

CLAUSE	OBLIGATIONS	TIME FRAME	SPECIFIC CONSEQUENCES OF NON-COMPLIANCE
6.6 Facilities for Staff and Labour	a) Provide and maintain all necessary accommodation and welfare facilities for the Contractor's Personnel. b) Provide facilities for the Employer's Personnel as stated in the Employer's Requirements. c) Not permit any of the Contractor's Personnel to maintain any temporary or permanent living quarters within the structures forming part of the Permanent Works.	None.	None.
6.7 Health and Safety	a) Take all reasonable precautions to maintain the health and safety of the Contractor's Personnel. b) Ensure that medical staff, first aid facilities, sick bay and ambulance service are available at all times, and that suitable arrangements are made for all necessary welfare and hygiene requirements, and for the prevention of epidemics. c) Appoint a qualified accident prevention officer and whatever is required by this person to exercise this responsibility and authority. d) Send to the Employer details of any accident. e) Maintain records and make reports concerning health, safety, welfare and damage to property.	a) None. b) None. c) None. d) As soon as practicable after its occurrence. e) None.	None.
6.8 Contractor's Superintendence	Provide all necessary superintendence to plan, arrange, direct, manage, inspect and test the work.	None.	None.
6.10 Records of Contractor's Personnel and Equipment	Submit to the Employer, details showing the number of each class of Contractor's Personnel and of each type of Contractor's Equipment on the Site.	Each calendar month.	None.
6.11 Disorderly Conduct	Take all reasonable precautions to prevent any unlawful, riotous or disorderly conduct by, or amongst the Contractor's Personnel.	None.	None.

THE OBLIGATIONS OF THE CONTRACTOR

Silver Book

THE OBLIGATIONS OF THE CONTRACTOR (continued)

CLAUSE	OBLIGATIONS	TIME FRAME	SPECIFIC CONSEQUENCES OF NON-COMPLIANCE
7 Plant, Materials and Workmanship			
7.1 Manner of Execution	Carry out the manufacture of Plant, the production and manufacture of Materials and all other execution of the Works.	None.	None.
7.2 Samples	Submit samples of Materials and relevant information to the Employer for review.	None.	None
7.3 Inspection	a) Give the Employer's Personnel full opportunity to carry out inspections. b) Give notice to the Employer to inspect.	a) None. b) Whenever any work is ready and before it is covered up, put out of sight or packaged for storage or transport.	a) None. b) Uncover the work, reinstate and make good at the Contractor's cost.
7.4 Testing	a) Provide everything necessary to carry out the specified tests. b) Agree with the Employer, the time and place for the testing. c) If the Contractor suffers delay and/or incurs Costs, give notice to the Employer. d) Forward to the Employer certified reports of the tests.	a) None. b) None. c) As soon as practicable and not later than 28 days after the Contractor became aware, or should have become aware of the event or circumstance (Sub-Clause 20.1). d) Promptly.	a) None. b) None. c) Loss of entitlement to an extension to the Time for Completion and additional payment (Sub-Clause 20.1). d) None.
7.5 Rejection	Make good defects notified by the Employer.	Promptly.	None.

CLAUSE	OBLIGATIONS	TIME FRAME	SPECIFIC CONSEQUENCES OF NON-COMPLIANCE
7.6 Remedial Work	Comply with the instructions of the Employer with regard to remedial work.	None.	Contractor shall pay costs incurred by the Employer in engaging other persons to carry out the work.
7.8 Royalties	Pay all royalties, rents and other payments for natural Materials obtained from outside the Site and disposal of surplus materials.	None.	None.
8 Commencement, Delays and Suspension			
8.1 Commencement of Works	Commence the execution of the Works and proceed with the Works with due expedition and without delay.	As soon as is reasonably practicable after the Commencement Date.	None.
8.2 Time for Completion	Complete the whole of the Works and each Section within the times specified in the Contract.	None.	Contractor shall pay delay damages to the Employer (Sub-Clause 8.7).
8.3 Programme	a) Submit a time programme. b) Submit a revised programme. c) Proceed in accordance with the programme. d) Give notice to the Employer of specific probable future events or circumstances which may adversely affect or delay the execution of the Works. e) Submit a revised programme on receiving a notice from the Employer that a programme fails to comply with the Contract, or to be consistent with actual progress.	a) Within 28 days of the Commencement Date. b) Whenever the previous programme is inconsistent with actual progress or with the Contractor's obligations. c) None. d) Promptly. e) None.	None.

THE OBLIGATIONS OF THE CONTRACTOR (continued)

CLAUSE	OBLIGATIONS	TIME FRAME	SPECIFIC CONSEQUENCES OF NON-COMPLIANCE
8.4 Extension of Time for Completion	Give notice to the Employer if the Contractor considers himself to be entitled to an extension of the Time for Completion.	As soon as practicable and not later than 28 days after the Contractor became aware, or should have become aware of the event or circumstance (Sub-Clause 20.1).	Loss of entitlement to an extension to the Time for Completion (Sub-Clause 20.1).
8.6 Rate of Progress	Adopt revised methods in order to expedite progress and complete within the Time for Completion.	None.	None.
8.7 Delay Damages	Pay delay damages in the case of failure to comply with the Time for Completion.	None.	None.
8.8 Suspension of Work	Protect, store and secure such part or all of the Works in the case of an instruction to suspend the Works.	None.	None.
8.9 Consequences of Suspension	Give notice to the Employer if the Contractor suffers delay and/or or incurs Cost from complying with the Employer's instructions under this clause or from resuming work.	As soon as practicable and not later than 28 days after the Contractor became aware, or should have become aware of the event or circumstance (Sub-Clause 20.1).	Loss of entitlement to an extension to the Time for Completion and additional payment (Sub-Clause 20.1).
8.12 Resumption of Work	a) Jointly examine the Works and the Plant and Materials affected by the suspension with the Employer. b) Make good any deterioration, defect or loss.	None.	None.

CLAUSE	OBLIGATIONS	TIME FRAME	SPECIFIC CONSEQUENCES OF NON-COMPLIANCE
9 Tests on Completion			
9.1 Contractor's Obligations	a) Carry out the Tests on Completion. b) Give the Employer notice of the date after which the Contractor will be ready to carry out each of the Tests on Completion. c) In the case of trial operation, give the Employer notice that the Works are ready for any other Tests on Completion. d) Submit a certified report of the results of the Tests to the Employer.	a) After providing the documents in accordance with Sub-Clauses 5.6 and 5.7. b) Not less than 21 days. c) When the Works are operating under stable conditions. d) As soon as the Works, or a Section have passed the Tests on Completion.	None.
9.2 Delayed Tests	Carry out the Tests if the Employer gives notice of undue delay.	Within 21 days of the Employer's notice.	The Employers Personnel may proceed with the tests at the Contractor's Cost.
10 Employer's Taking Over			
10.3 Interference with Tests on Completion	a) In the case of prevention from carrying out the tests, carry out any outstanding Tests on Completion. b) If the Contractor suffers delay and/or incurs Cost as a result of delay in carrying out the Tests on Completion, give notice to the Employer.	a) As soon as practicable. b) As soon as practicable and not later than 28 days after the Contractor became aware, or should have become aware of the event or circumstance (Sub-Clause 20.1).	a) None. b) Loss of entitlement to an extension to the Time for Completion and additional payment (Sub-Clause 20.1).

THE OBLIGATIONS OF THE CONTRACTOR

Silver Book

THE OBLIGATIONS OF THE CONTRACTOR (continued)

CLAUSE	OBLIGATIONS	TIME FRAME	SPECIFIC CONSEQUENCES OF NON-COMPLIANCE
11 Defects Liability			
11.1 Completion of Outstanding Work and Remedying Defects	a) Complete any work which is outstanding on the date stated in a Taking-Over Certificate. b) Execute all work required to remedy defects or damage.	a) Within such reasonable time as is instructed by the Employer. b) On or before the expiry date of the Defects Notification Period.	a) The Employer may carry out the work himself, at the Contractor's Cost (Sub-Clause 11.4). b) A reduction in the Contract Price may be made (Sub-Clause 11.4).
11.8 Contractor to Search	If required by the Employer, search for the cause of any defect.	None.	None.
11.11 Clearance of Site	Remove any remaining Contractor's Equipment, surplus material, wreckage, rubbish and Temporary Works from the Site.	Within 28 days of receipt of the Performance Certificate.	The Employer may sell or otherwise dispose of any remaining items and the Employer is entitled to recover the costs of disposal from the Contractor.

CLAUSE	OBLIGATIONS	TIME FRAME	SPECIFIC CONSEQUENCES OF NON-COMPLIANCE
12 Tests after Completion			
12.1 Procedure for Tests after Completion	a) Provide plant, equipment and suitably qualified staff to carry out the tests. b) Carry out the tests in the presence of the Employer's Personnel, if requested. c) Compile and evaluate the results of the Tests after Completion.	As soon as reasonably practicable after the Works or Section have been taken over by the Employer.	None.
12.2 Delayed Tests	If the Contractor incurs Cost as a result of unreasonable delay by the Employer, give notice to the Employer.	As soon as practicable and not later than 28 days after the Contractor became aware, or should have become aware of the event or circumstance (Sub-Clause 20.1).	Loss of entitlement to additional payment (Sub-Clause 20.1).
12.3 Retesting	If the Works or a Section fail to pass the Tests after Completion, execute all work required to remedy defects or damage.	On or before the expiry date of the Defects Notification Period.	A reduction in the Contract Price may be made (Sub-Clause 11.4).
12.4 Failure to Pass Tests after Completion	If the Contractor incurs additional Cost as a result of any unreasonable delay by the Employer, give notice to the Employer.	As soon as practicable and not later than 28 days after the Contractor became aware, or should have become aware of the event or circumstance (Sub-Clause 20.1).	Loss of entitlement to additional payment (Sub-Clause 20.1).
13 Variations and Adjustments			
13.1 Right to Vary	Execute and be bound by each Variation.	None.	None.
13.3 Variation Procedure	a) Respond in writing to a request for a proposal. b) Not delay any work whilst awaiting a response. c) Acknowledge receipt of Variation instructions.	a) As soon as practicable. b) None. c) None.	None.

Silver Book

THE OBLIGATIONS OF THE CONTRACTOR (continued)

CLAUSE	OBLIGATIONS	TIME FRAME	SPECIFIC CONSEQUENCES OF NON-COMPLIANCE
13.5 Provisional Sums	Produce quotations, invoices, vouchers and accounts or receipts in substantiation of the amounts paid to nominated Subcontractors.	When required by the Employer.	None.
13.6 Daywork	a) Submit quotations for Goods to the Employer. b) Submit invoices, vouchers and accounts or receipts for Goods. c) Deliver to the Employer, statements which include the details of the resources used in executing the previous day's work. d) Submit priced statements of these resources.	a) Before ordering Goods for the work to be executed on a Daywork basis. b) When applying for payment. c) Each day. d) Prior to their inclusion in the next Statement under Sub-Clause 14.3.	None.
13.7 Adjustments for Changes in Legislation	If the Contractor suffers (or will suffer) delay and/or incurs (or will occur) additional Cost, give notice to the Employer.	As soon as practicable and not later than 28 days after the Contractor became aware, or should have become aware of the event or circumstance (Sub-Clause 20.1).	Loss of entitlement to an extension to the Time for Completion and additional payment (Sub-Clause 20.1).
14 Contract Price and Payment			
14.1 The Contract Price	Pay all taxes, duties and fees required to be paid under the Contract.	None.	None.
14.2 Advance Payment	a) Submit an advance payment guarantee. b) Extend the validity of the guarantee until the advance payment has been repaid.	None.	a) Employer is not obligated to make the advance payment. b) None.

CLAUSE	OBLIGATIONS	TIME FRAME	SPECIFIC CONSEQUENCES OF NON-COMPLIANCE
14.3 Application for Interim Payment	Submit, in six copies, a Statement showing in detail the amounts to which the Contractor considers himself to be entitled.	After the period of payment stated in the Contract, or after the end of each month.	No obligation on the Employer to make payment (Sub-Clause 14.6).
14.4 Schedule of Payments	In the case that the Contract does not include a schedule of payments, submit non-binding estimates of the payments expected to become due.	a) First estimate within 42 days after the Commencement Date. b) Revised estimates at quarterly intervals.	None.
14.10 Statement at Completion	Submit a Statement at completion.	Within 84 days after receiving the Taking-Over Certificate for the Works.	No obligation on the Employer to make payment (Sub-Clause 14.6).
14.11 Application for Final Payment Certificate	a) Submit a draft final statement. b) Submit such further information as the Employer may reasonably require. c) Prepare and submit the Final Statement as agreed with the Employer.	a) Within 56 days after receiving the Performance Certificate. b) None. c) None.	a) No obligation on the Employer to make payment (Sub-Clause 14.6). b) None. c) None.
14.12 Discharge	Submit a written discharge.	When submitting the Final Statement.	None.
15 Termination by Employer			
15.2 Termination by Employer	a) In the case of a notice of termination being served, leave the Site and deliver any required Goods, Contractor's Documents and other design documents to the Employer. b) Use best efforts to comply with any reasonable instructions included in the notice. c) Arrange for the removal of Equipment and Temporary Works.	a) None. b) Immediately. c) Promptly.	a) None. b) None. c) Items may be sold by the Employer.

THE OBLIGATIONS OF THE CONTRACTOR (continued)

CLAUSE	OBLIGATIONS	TIME FRAME	SPECIFIC CONSEQUENCES OF NON-COMPLIANCE
15.5 Employer's Entitlement to Termination	In the case of a notice of termination, cease all further work, hand over Contractor's Documents, Plant, Materials and other work and remove all other Goods from the Site (Sub-Clause 16.3).	28 days from the Employer's notice or the return of the Performance Security, whichever is the later.	None.
16 Suspension and Termination by Contractor			
16.1 Contractor's Entitlement to Suspend Work	a) Give notice if the Contractor intends to suspend work or reduce the rate of work. b) Resume normal working when the Employer's obligations have been met. c) If the Contractor suffers delay and/or incurs Cost, give notice to the Employer.	a) 21 days before the intended suspension or reduction in the rate of work. b) As soon as is reasonably practicable. c) As soon as is practicable and not later than 28 days after the Contractor became aware, or should have become aware of the event or circumstance (Sub-Clause 20.1).	a) None. b) None. c) Loss of entitlement to an extension to the Time for Completion and additional payment (Sub-Clause 20.1).
16.2 Termination by Contractor	Give notice of intention to terminate.	14 days before the intended termination date.	None.
16.3 Cessation of Work and Removal of Contractor's Equipment	In the case of a notice of termination, cease all further work, hand over Contractor's Documents, Plant, Materials and other work and remove all other Goods from the Site.	After the notice has taken effect.	None.

CLAUSE	OBLIGATIONS	TIME FRAME	SPECIFIC CONSEQUENCES OF NON-COMPLIANCE
17 Risk and Responsibility			
17.1 Indemnities	Indemnify and hold harmless the Employer, the Employer's Personnel and their respective agents against and from all claims, damages, losses and expenses in respect of bodily injury, sickness, disease, death, damage to or of loss of property by reason of the Contractor's design, the execution and completion of the Works.	None.	None.
17.2 Contractor's Care of the Works	a) Take full responsibility for the care of the Works and Goods. b) Take responsibility for the care of any work which is outstanding on the date stated in a Taking-Over Certificate. c) Rectify loss or damage if any loss or damage happens to the Works, Goods or Contractor's Documents.	a) From the Commencement Date until the Taking-Over Certificate is issued. b) Until the outstanding work has been completed. c) None.	None.
17.4 Consequences of Employer's Risks	a) Give notice in the case of an Employer's risk event that results in loss or damage. b) Rectify the loss or damage as required by the Employer. c) If the Contractor suffers delay and/or incurs Cost from rectifying loss or damage, give further notice to the Employer.	a) Promptly. b) None. c) As soon as practicable and not later than 28 days after the Contractor became aware, or should have become aware of the event or circumstance (Sub-Clause 20.1).	a) None. b) None. c) Loss of entitlement to an extension to the Time for Completion and additional payment (Sub-Clause 20.1).
17.5 Intellectual and Industrial Property Rights	a) Indemnify and hold the Employer harmless against and from any other claim which arises out of, or in relation to the Contactor's design, manufacture, construction or execution of the Works, the use of Contractor's equipment or the proper use of the Works. b) If requested by the Employer, assist in contesting the claim. c) Not make any admission which might be prejudicial to the Employer.	a) None. b) None. c) None.	a) None. b) None. c) None.

Silver Book

THE OBLIGATIONS OF THE CONTRACTOR (continued)

CLAUSE	OBLIGATIONS	TIME FRAME	SPECIFIC CONSEQUENCES OF NON-COMPLIANCE
18 Insurance			
18.1 General Requirements for Insurance	a) Wherever the Contractor is the insuring Party, effect and maintain the insurances in terms consistent with any terms agreed by the Parties before the signing of the Contract Agreement. b) Act under the policy on behalf of any additional joint insured parties. c) Submit evidence to the Employer that the insurances have been effected and copies of the policies. d) Submit evidence of payment of premiums. e) Inform the insurers of any relevant changes to the execution of the Works and ensure that insurance is maintained. f) Not make any material alteration to the terms of any insurance without approval of the Employer.	a) None. b) None. c) Within the periods stated in the Particular Conditions. d) Upon payment of premium. e) As appropriate. f) None.	Employer may effect the insurances and recover the cost from the Contractor.
18.4 Insurance for Contractor's Personnel	Effect and maintain insurance against injury, sickness, disease or death of any person employed by the Contractor, or any other of the Contractor's Personnel.	From the time that these personnel are assisting in the execution of the Works.	Employer may effect the insurances and recover the Cost from the Contractor (Sub-Clause 18.1).
19 Force Majeure			
19.2 Notice of Force Majeure	Give notice in the case that the Contractor is, or will be prevented from performing any of its obligations under the Contract by Force Majeure.	Within 14 days after the Contractor became aware, or should have become aware of the relevant event or circumstance constituting Force Majeure.	Contractor shall not be excused performance of the obligations.

CLAUSE	OBLIGATIONS	TIME FRAME	SPECIFIC CONSEQUENCES OF NON-COMPLIANCE
19.3 Duty to Minimise Delay	c) Use all reasonable endeavours to minimise any delay in the performance of the Contract as a result of Force Majeure. d) Give notice when the effects of the Force Majeure cease.	None.	None.
20 Claims, Disputes and Arbitration			
20.1 Contractor's Claims	a) Give notice if the Contractor considers himself to be entitled to any extension of the Time for Completion and/or any additional payment. b) Submit any other notices which are required by the Contract and supporting particulars of the claim. c) Keep such contemporary records as may be necessary to substantiate any claim and permit the Employer to inspect all the records. d) Send to the Employer, a fully detailed claim. e) Send further interim claims if the event or circumstance giving rise to the claim has a continuing effect. f) Send a final claim.	a) As soon as practicable and not later than 28 days after the Contractor became aware, or should have become aware of the event or circumstance. b) None. c) None. d) Within 42 days after the Contractor became aware (or should have become aware) of the event or circumstance giving rise to the claim. e) At monthly intervals. f) Within 28 days after the end of the effects resulting from the event or circumstance.	a,b,d) Loss of entitlement to an extension to the Time for Completion and additional payment. c,e,f) The Employer will take account of the extent to which the failure has prevented or prejudiced proper investigation of the claim.
20.2 Appointment of the Dispute Adjudication Board	d) Jointly with the Employer, appoint the Dispute Ajudication Board (DAB). e) Mutually agree the terms of remuneration for the DAB. f) Not to act alone in the termination of any member of the DAB.	a) Within 28 days after either Party gives notice of intention to refer a dispute to a DAB. b) None. c) None.	a) The appointing entity or official named in the Particular Conditions shall appoint (Sub-Clause 20.3). b) None. c) None.

THE OBLIGATIONS OF THE CONTRACTOR (continued)

CLAUSE	OBLIGATIONS	TIME FRAME	SPECIFIC CONSEQUENCES OF NON-COMPLIANCE
20.4 Obtaining Dispute Adjudication Board's Decision	c) Make available to the DAB, additional information, access to the Site and appropriate facilities as the DAB may require. d) Give effect to a DAB decision unless and until it is revised in an amicable settlement or an arbitral award. e) Continue to proceed with the Works in accordance with the Contract.	c) Promptly. d) Promptly. e) None.	a) None. b&c) The matter may be referred to arbitration (Sub-Clause 20.7).
20.5 Amicable Settlement	In the case of a notice of dissatisfaction being issued, attempt to settle the dispute amicably.	Within 56 days of the notice.	Arbitration may be commenced.

GENERAL CONDITIONS OF DISPUTE ADJUDICATION AGREEMENT

2 General Provisions	Give notice to the DAB Member that the Dispute Adjudication Agreement has taken effect.	Upon all parties signing the Dispute Adjudication Agreement.	None.
5 General Obligations of the Employer and the Contractor	Not to request advice from, or consult with the Member regarding the Contract, otherwise than in the normal course of the DAB's activities.	None.	None.

CLAUSE	OBLIGATIONS	TIME FRAME	SPECIFIC CONSEQUENCES OF NON-COMPLIANCE
6 Payment	a) Pay the DAB an advance payment for 25% of the estimated fees and expenses. b) Pay the DAB's invoices. c) Apply to the Employer for reimbursement of one-half of the DAB invoices by way of the Statements.	a) Upon receipt of the invoice. b) Within 28 days of receipt. c) None.	a) DAB member not obliged to engage in activities under the DAB Agreement. b) DAB member not obliged to render its decision. a&b) Employer entitled to pay fees and recover reimbursement of fees, plus financing charges from the Contractor. a&b) DAB Member may suspend services or resign the appointment. c) The Employer is not obliged to reimburse the Contractor.
Annex – Procedural Rules			
1.	a) Jointly with the Employer, furnish to each member of the DAB one copy of all documents which the DAB may request. b) Copy the Employer on all communications between the DAB and the Contractor.	None.	None.

THE OBLIGATIONS OF THE DISPUTE ADJUDICATION BOARD

CLAUSE	OBLIGATIONS	TIME FRAME	SPECIFIC CONSEQUENCES OF NON-COMPLIANCE
GENERAL CONDITIONS			
20 Claims, Disputes and Arbitration			
20.4 Obtaining Dispute Adjudication Board's Decision	Give a reasoned decision on any dispute referred to the DAB.	Within 84 days after receiving a dispute reference or the advance payment, whichever is the later.	Either Party may give a notice of dissatisfaction and commence arbitration.
GENERAL CONDITIONS OF DISPUTE ADJUDICATION AGREEMENT			
3 Warranties	a) Be impartial and independent of the Employer, the Contractor and the Employer's Representative. b) Disclose to the Parties and to the Other Members, any fact or circumstance that might appear inconsistent with his/her warranty and agreement of impartiality and independence.	a) None. b) Promptly.	None.

CLAUSE	OBLIGATIONS	TIME FRAME	SPECIFIC CONSEQUENCES OF NON-COMPLIANCE
4 General Obligations of the Member	a) Have no interest, financial or otherwise in the Parties, nor any financial interest in the Contract. b) Not previously have been employed as a consultant or otherwise by the Parties, except as disclosed in writing. c) Disclose in writing to the Parties and the Other Members, any professional or personal relationships with any director, officer or employee of the Parties and any previous involvement in the overall project of which the Contract forms part. d) Not, for the duration of the Dispute Adjudication Agreement, to be employed as a consultant or otherwise by the Parties, except as may be agreed in writing. e) Comply with the procedural rules and with Sub-Clause 20.4 of the Conditions of Contract. f) Not to give advice to the Parties, the Employer's Personnel or the Contractor's Personnel concerning the conduct of the Contract, other than in accordance with the procedural rules. g) Not to enter into discussions or make any agreement with the Employer or the Contractor regarding employment by any of them after ceasing to act under the Dispute Adjudication Agreement. h) Ensure his/her availability for all site visits and hearings as are necessary. i) Treat the details of the Contract and all the DAB's activities and hearings as private and confidential.	None.	None.
6 Payment	a) Submit an invoice for an advance of 25% of the estimated total amount of daily fees and the estimated total of expenses that will be incurred. b) Submit invoices for the balance of daily fees and expenses.	a) Immediately after the Dispute Adjudication Agreement takes effect and before engaging in any activities. b) None.	None.

THE OBLIGATIONS OF THE DISPUTE ADJUDICATION BOARD (continued)

CLAUSE	OBLIGATIONS	TIME FRAME	SPECIFIC CONSEQUENCES OF NON-COMPLIANCE
Annex – Procedural Rules			
1.	Copy all communications between the DAB and the Employer or the Contractor to the other Party.	None.	None.
2(a).	a) Act fairly and impartially as between the Parties. b) Give each of the Parties a reasonable opportunity of putting his case and responding to the other's case.	None.	None.
2(b).	Adopt procedures suitable to the dispute, avoiding unnecessary delay or expense.	None.	None.
3.	In the case of a hearing on the dispute, decide on the date and place for the hearing.	None.	None.
6.	a) Not express any opinions during any hearing concerning the merits of any arguments advanced by the Parties. b) Make and give a decision in accordance with Sub-Clause 20.4, or as otherwise agreed by the Employer and the Contractor in writing. c) If the DAB comprises three persons: I. Convene in private after a hearing. II. Endeavour to reach a unanimous decision.	None.	None.

Chapter 6
The Gold Book
Conditions of Contract for Design, Build and Operate Projects, First Edition 2008

The FIDIC Contracts: Obligations of the Parties, First Edition. Andy Hewitt.
© 2014 John Wiley & Sons, Ltd. Published 2014 by John Wiley & Sons, Ltd.

THE OBLIGATIONS OF THE EMPLOYER

CLAUSE	OBLIGATIONS	TIME FRAME	SPECIFIC CONSEQUENCES OF NON-COMPLIANCE
GENERAL CONDITIONS			
1 General Provisions			
1.6 Contract Agreement	Enter into a Contract Agreement.	Within 28 days after the Contractor receives the Letter of Acceptance unless agreed otherwise.	None.
1.7 Operating License	Issue the Operating License or equivalent legal authorisation to enable the Contractor to operate and maintain the Works.	Together with the Letter of Acceptance.	None.
1.9 Care and Supply of Documents	In the case of an error in a document, give Notice to the Contractor.	Promptly.	None.
1.11 Employer's Use of Contractor's Documents	Not to, without the Contactor's consent, use, copy or communicate to a third party the Contractor's Documents.	None.	None.
1.13 Confidential Details	Treat all information designated as confidential by the Contractor as confidential.	None.	None.
1.14 Compliance with Laws	Obtain the planning, zoning or similar permission for the Permanent Works and the Operation Service and any other permissions described in the Employer's Requirements as having been (or being) obtained by the Employer.	None.	None.

THE OBLIGATIONS OF THE EMPLOYER (continued)

CLAUSE	OBLIGATIONS	TIME FRAME	SPECIFIC CONSEQUENCES OF NON-COMPLIANCE
2 The Employer			
2.1 Right of Access to the Site	a) Give the Contractor right of access to and possession of all parts of the Site. b) Give the Contractor possession of any foundation, structure, plant or means of access if required.	a) Within the time (or times) stated in the Contract Data. b) In the time and manner stated in the Contract Data.	Contractor shall be entitled to an extension of time and payment of Cost plus reasonable profit.
2.2 Permits, Licenses or Approvals	Provide reasonable assistance to the Contractor at the request of the Contractor: a) To obtain copies of the Laws of the Country which are relevant but not readily available and b) To obtain permits, licences or approvals required by the Laws of the Country.	None.	None.
2.3 Employer's Personnel	Ensure that the Employer's Personnel and the Employer's other contractors cooperate with the Contractor and take actions similar to those which the Contractor is required to take under Sub-Clause 4.8 [Safety Procedures] and under Sub-Clause 4.18 [Protection of the Environment].	None.	None.
2.4 Employer's Financial Arrangements	a) In the case that the Employer intends to make material changes to financial arrangements, give notice and detailed particulars to the Contractor. b) Submit reasonable evidence that financial arrangements have been made and are being maintained, which will enable the Employer to pay the Contract Price.	a) None. b) Within 28 days after receiving a request from the Contractor.	Contractor entitled to suspend work, reduce the rate of work and to an extension of time and additional payment as a result of such actions (Sub-Clause 16.1).

CLAUSE	OBLIGATIONS	TIME FRAME	SPECIFIC CONSEQUENCES OF NON-COMPLIANCE
3 The Employer's Representative			
3.1 Employer's Representative's Duties and Authority	a) Appoint a suitably qualified and experienced Employer's Representative to carry out the duties assigned to him in the Contract. b) Not to impose further constraints on the Employer's Representative's authority except as agreed with the Contractor.	a) Prior to the signing of the Contract. b) None.	None.
3.4 Replacement of the Employer's Representative	a) Give Notice to the Contractor of the name, address and relevant experience of the intended replacement Employer's Representative. b) Not to replace the Employer's Representative with a person against whom the Contractor raises reasonable objection, by Notice to the Employer, with supporting particulars.	a) Not less than 42 days before the intended date of replacement. b) None.	None.
3.5 Determinations	Give effect to Notices of agreement or determination given by the Employer's Representative.	None.	None.
4 The Contractor			
4.2 Performance Security	a) Cooperate with the Contractor to agree the entity, country (or other jurisdiction) for the issue of the Performance Security. b) Cooperate with the Contractor to agree the form of Performance Security if not in the form annexed to the tender documents. c) Not make a claim under the Performance Security, except for amounts to which the Employer is entitled under the Contract (as listed). d) Return the Performance Security to the Contractor.	a) None. b) None. c) None. d) Within 21 days after receiving the Contract Completion Certificate.	None.

THE OBLIGATIONS OF THE EMPLOYER (continued)

CLAUSE	OBLIGATIONS	TIME FRAME	SPECIFIC CONSEQUENCES OF NON-COMPLIANCE
4.10 Site Data	a) Make available to the Contractor all relevant data in the Employer's possession on sub-surface, hydrological and climatic conditions at the Site, including environmental aspects. b) Make available to the Contractor all such data which come into the Employer's possession after the Base Date.	a) Prior to the Base Date. b) None.	None.
4.19 Electricity, Water and Gas	a) Allow the Contractor to use such supplies of electricity, water, gas and other services as may be available on the Site and of which details are given in the Employer's Requirements.	None.	None.
4.20 Employer's Equipment and Free-issue Materials	a) Make the Employer's Equipment (if any) available for the use of the Contractor in the execution of the Works in accordance with the details, arrangements and prices stated in the Employer's Requirements. b) Supply, free of charge, the 'free-issue materials' (if any) in accordance with the details stated in the Employer's Requirements. c) Rectify any notified shortage, defect or default in the free-issue materials.	a) None. b) As specified in the Contract. c) Immediately.	None.
4.24 Fossils	Take care of and exercise authority over fossils, coins, articles of value or antiquity, structures and other remains or items of geological or archaeological interest found on the Site.	None.	None.
4.25 Changes in the Contractor's Financial Situation	In the case of a Notice from the Contractor, advise the Contractor of what action the Employer intends to take and what action he requires the Contractor to take.	Within 28 days of receiving the Notice.	None.

CLAUSE	OBLIGATIONS	TIME FRAME	SPECIFIC CONSEQUENCES OF NON-COMPLIANCE
10 Operation Service			
10.3 Independent Compliance Audit	a) Jointly with the Contractor, appoint the Auditing Body to carry out an audit during the Operation Service. b) Cooperate with the Auditing Body and give due regard to matters raised in each report issued by the Auditing body.	a) At least 182 days prior to the commencement of the Operation Service. b) None.	a) The matter may be referred to the DAB and the DAB shall make the appointment. b) None.
10.4 Delivery of Raw Materials	Supply and deliver free issue raw materials, fuel, consumables and other such items specified in the Employer's Requirements.	In accordance with the agreed delivery programme.	The Contractor shall be entitled to recover Cost plus Profit incurred as a result.
10.5 Training	a) Provide the training facilities for the Contractor to carry out the training of the Employer's Personnel. b) Nominate and select suitable personnel for training.	None.	None.
10.6 Delays and Interruptions during the Operation Service	a) In the case of delay or interruptions caused by the Employer, compensate the Contractor for any cost and losses. b) Pay the amount due.	a) None. b) Within the next payment due to the Contractor.	None.
10.7 Failure to Reach Production Outputs	a) In the event that the Contractor fails to achieve production outputs, jointly with the Contractor establish the cause of such failure. b) If the cause of the failure lies with the Employer, give written instructions to the Contractor of the measures which are required to be undertaken by the Contractor. c) In the case that the Contractor suffers additional cost, pay the Contractor his Cost plus Profit.	None.	None.

THE OBLIGATIONS OF THE EMPLOYER (continued)

CLAUSE	OBLIGATIONS	TIME FRAME	SPECIFIC CONSEQUENCES OF NON-COMPLIANCE
13 Variations and Adjustments			
13.1 Right to Vary	In the case that the Employer wishes to instruct a Variation during the Operation Service Period, give the Contractor written details of the requirements.	None.	None.
14 Contract Price and Payment			
14.2 Advance Payment	Make the advance payment as an interest-free loan for mobilisation and design.	When the Contractor submits a guarantee.	Contractor entitled to suspend work, reduce the rate of work and to an extension of time and additional payment as a result of such actions (Sub-Clause 16.1).
14.8 Payments	a) Pay the advance payment. b) Pay the amount due in respect of each Statement. c) Pay the amount due in respect of the Final Payment Certificates.	a) Within 21 days after receiving the documents in accordance with Sub-Clause 4.2 [Performance Security], Sub-Clause 14.2 [Advance Payment] and the Payment Certificate for the advance payment. b) Within 56 days after receipt of the Statement and supporting documents by the Employer's Representative. c) Within 56 days after receipt of each of the Final Payment Certificates.	1) Contractor entitled to suspend work, reduce the rate of work and to an extension of time and additional payment as a result of such actions (Sub-Clause 16.1). 2) Payment of financing charges to the Contractor (Sub-Clause 14.9).

CLAUSE	OBLIGATIONS	TIME FRAME	SPECIFIC CONSEQUENCES OF NON-COMPLIANCE
14.12 Issue of Final Payment Certificate Design-Build	Pay the Contactor the amount due on the Final Payment Certificate Design-Build.	Within 56 days after receipt of the Certificate.	Payment of financing charges to the Contractor (Sub-Clause 14.9).
14.15 Issue of Final Payment Certificate Operation Service	Pay the Contactor the amount due on the Final Payment Certificate Operation Service.	Within 56 days after receipt of the Certificate.	Payment of financing charges to the Contractor (Sub-Clause 14.9).
14.17 Currencies of Payment	Pay the Contractor in the currency or currencies named in the Contract Agreement.	None.	None.
14.18 Asset Replacement Fund	Authorise release of funds from the Asset Replacement Fund in accordance with the amounts certified by the Employer's Representative.	None.	None.
15 Termination by Employer			
15.2 Termination for Contractor's Default	a) Give Notice of intention to terminate the Contract. b) Give Notice of release of the Contractor's Equipment and Temporary Works.	a) 14 days prior to termination date, or immediately in the case of the Contractor becoming bankrupt or gives or offers bribes or gratuities (or similar as defined in the clause). b) On completion of the Works.	None.
15.4 Payment after Termination for Contractor's Default	Pay the balance due to the Contractor after recovering any losses, damages and extra costs.	None.	None.

THE OBLIGATIONS OF THE EMPLOYER (continued)

CLAUSE	OBLIGATIONS	TIME FRAME	SPECIFIC CONSEQUENCES OF NON-COMPLIANCE
15.5 Termination for Employer's Convenience	a) Give Notice of intention to terminate the Contract. b) Make arrangement to return the Performance Security. c) Return the Contractor's Documents to the Contractor. d) Not to terminate the Contract in order to execute the Works himself or to arrange for the Works to be executed by another contractor.	a) 28 days prior to the termination date. b) Immediately. c) Forthwith after issue of the Notice. d) None.	None.
16 Suspension and Termination by Contractor			
16.4 Payment on Termination	a) Return the Performance Security to the Contractor. b) Pay the Contractor in accordance with Sub-Clause 19.6 [Optional Termination, Payment and Release]. c) Pay to the Contractor the amount of any loss of profit or other loss or damage sustained by the Contractor as a result of the termination.	Promptly.	None.
17 Risk Allocation			
17.10 Indemnities by the Employer	Indemnify and hold harmless the Contractor, the Contractor's Personnel and their respective agents against and from all claims, damages, losses and expenses in respect of bodily injury, disease or death which is attributable to any negligence, wilful act or breach of the Contract by the Employer, the Employer's personnel or agents and the Employer's Risks as set out in Sub-Clauses 17.1 and 17.3.	None.	None.

CLAUSE	OBLIGATIONS	TIME FRAME	SPECIFIC CONSEQUENCES OF NON-COMPLIANCE
17.12 Risk of Infringement of Intellectual and Industrial Property Rights	a) Indemnify and hold the Contractor harmless against and from any claim, which is or was, an unavoidable result of the Contractor's compliance with the Employer's Requirements, or as a result of any Works being used by the Employer. b) If requested by the Contractor, assist in contesting the claim. c) Not make any admission which might be prejudicial to the Contractor.	None.	None.
18 Exceptional Risks			
18.2 Notice of an Exceptional Event	In the case of an Exceptional Event that prevents the Employer from performing the Employer's obligations, give Notice to the Contractor and specify the affected obligations.	Within 14 days after the Employer became aware, or should have become aware of the event or circumstances.	None.
18.3 Duty to Minimise Delay	a) Use all reasonable endeavours to minimise any delay. b) Give Notice to the Contractor when the effects of the Exceptional Event cease.	None.	None.
19 Insurance			
19.1 General Requirements	Not to unreasonably withhold approval of the insurers and the terms of the Contractor's insurances.	None.	None.
20 Claims, Disputes and Arbitration			
20.2 Employer's Claims	In the case that the Employer considers himself to be entitled to any payment, give Notice and particulars of the claim to the Contractor (alternatively, such Notice may be issued by the Employer's Representative).	As soon as practicable after the Employer becomes aware, or should have become aware of the event or circumstances giving rise to the claim.	None.

THE OBLIGATIONS OF THE EMPLOYER

Gold Book

THE OBLIGATIONS OF THE EMPLOYER (continued)

CLAUSE	OBLIGATIONS	TIME FRAME	SPECIFIC CONSEQUENCES OF NON-COMPLIANCE
20.3 Appointment of the Dispute Adjudication Board	a) Jointly, with the Contractor, appoint the Dispute Adjudication Board (DAB). b) Not to act alone in the termination of any member of the DAB.	a) By the date stated in the Contract Data. b) None.	a) The appointing entity or official named in the Contract Data shall appoint the DAB (Sub-Clause 20.4). b) None.
20.6 Obtaining Dispute Adjudication Board's Decision	a) Make available to the DAB additional information, access to the Site, and appropriate facilities as the DAB may require. b) Comply with a DAB decision. c) In the case of dissatisfaction with the DAB's decision, or if the dissatisfaction with the fact that the DAB has not given a decision within the prescribed time, give Notice to the Contractor with a copy to the chairman of the DAB.	a) Promptly. b) Promptly. c) Within 28 days of receiving the decision or the expiry of the period during which the decision should have been given.	a) None. b) The Contractor may refer the failure to arbitration (Sub-Clause 20.9). c) Arbitration may not be commenced and the DAB decision shall become final and binding.
20.7 Amicable Settlement	In the case of a Notice of dissatisfaction being issued, attempt to settle the dispute amicably.	Before the commencement of arbitration.	Arbitration may be commenced within 28 days of the Notice of dissatisfaction being given.
20.10 Disputes Arising during the Operation Service Period	a) Jointly agree on and appoint the Operation Service DAB. b) Jointly agree on and appoint a new Operation Service DAB.	a) At the time of the issue of the Commissioning Certificate. b) At the end of each five-year appointment term.	The appointing entity or official named in the Contract Data shall appoint the DAB.

CLAUSE	OBLIGATIONS	TIME FRAME	SPECIFIC CONSEQUENCES OF NON-COMPLIANCE
GENERAL CONDITIONS OF DISPUTE ADJUDICATION AGREEMENT			
5 General Obligations of the Employer and the Contractor	a) Not to request advice from or consultation with a DAB Member regarding the Contract otherwise than in the normal course of the DAB's activities. b) In the case that the Employer is the referring Party of a Dispute, provide appropriate security for a sum equivalent to the reasonable expenses to be incurred by the Member.	None.	None.
6 Payment	a) Pay one half of the DAB fees to the Contractor. b) In the case of the Contractor failing to pay the Member, pay the due fees.	a) In accordance with the Contract. b) None.	The DAB Member may suspend services or resign the appointment.
Procedural Rules for Dispute Adjudication Board Members			
2.	Jointly agree with the DAB and the Contractor, the timing and agenda for each Site visit.	None.	The timing and agenda will be decided by the DAB.
3.	a) Attend Site visits by the DAB. b) Co-ordinate the Site visits. c) Ensure the provision of conference facilities and secretarial and copying services.	None.	None.
4.	a) Jointly with the Contractor, furnish to each DAB member one copy of all documents which the DAB may request. b) Copy all communications with the DAB to the Contractor.	None.	None.

Gold Book

THE OBLIGATIONS OF THE CONTRACTOR

CLAUSE	OBLIGATIONS	TIME FRAME	SPECIFIC CONSEQUENCES OF NON-COMPLIANCE
GENERAL CONDITIONS			
1 General Provisions			
1.9 Care and Supply of Documents	a) Supply to the Employer's Representative six copies of each of the Contractor's Documents. b) Keep on the Site, a copy of the Contract, publications named in the Employer's Requirements, the Contractor's Documents and Variations and other communications given under the Contract. c) In the case of an error in a document, give Notice to the Employer.	a) None. b) None. c) Promptly.	None.
1.10 Errors in the Employer's Requirements	In the case that an error is found in the Employer's Requirements, give Notice to the Employer's Representative.	Immediately.	None.
1.12 Contractor's Use of Employer's Documents	Not to, without consent, copy, use or communicate the Employer's Requirements and other documents of the Employer to a third party, except as necessary.	None.	None.
1.13 Confidential Details	a) Disclose all such confidential and other information as the Employer's Representative may reasonably require in order to verify the Contractor's compliance with the Contract. b) Treat the details of the Contract as private and confidential. c) Not to publish, or permit to be published or disclose any particulars without the consent of the Employer.	None.	None.

CLAUSE	OBLIGATIONS	TIME FRAME	SPECIFIC CONSEQUENCES OF NON-COMPLIANCE
1.14 Compliance with Laws	a) Comply with applicable Laws. b) Give all Notices, pay all taxes, duties and fees, and obtain all permits, licences and approvals as required by the Laws. c) Comply with, give all Notices and pay all fees required by any license of the Site or the Works or the Operation Service.	None.	None.
1.15 Joint and Several Liability	a) In the case of a joint venture, consortium or other unincorporated grouping of two or more persons, notify the Employer of the leader. b) Not to alter the composition of the joint venture or legal status without the prior consent of the Employer.	None.	None.
2 The Employer			
2.1 Right of Access to the Site	Give Notice to the Employer's Representative if the Contractor suffers delay and/or incurs Cost as a result of failure to give right of access and possession of the Site.	As soon as practicable and not later than 28 days after the Contractor became aware, or should have become aware of the event or circumstance (Sub-Clause 20.1).	Loss of entitlement to an extension to the Time for Completion and additional payment (Sub-Clause 20.1).
3 The Employer's Representative			
3.3 Instructions of the Employer's Representative	a) Take instructions from the Employer's Representative, or from an assistant to whom the appropriate authority has been delegated. b) Comply with the instructions given by the Employer's Representative or delegated assistant. c) In the case that the Contractor considers that any instruction does not comply with applicable Laws or is technically impossible, notify the Employer's Representative.	a) None. b) None. c) Immediately.	None.
3.5 Determinations	Give effect to Notices of agreement or determinations given by the Employer's Representative.	None.	None.

THE OBLIGATIONS OF THE CONTRACTOR (continued)

CLAUSE	OBLIGATIONS	TIME FRAME	SPECIFIC CONSEQUENCES OF NON-COMPLIANCE
4 The Contractor			
4.1 Contractor's General Obligations	a) Design, execute and complete the Works and provide the Operation Service in accordance with the Contract and remedy any defects in the Works. b) Ensure that the Works remain fit for purpose during the Operation Service Period. c) Provide the required Plant and Contractor's Documents specified in the Contract and all Contractor's Personnel, Goods, consumables and other things and services, whether of a temporary or permanent nature. d) Be responsible for the adequacy, stability and safety of all Site operations and of all methods of construction. e) Submit details of the arrangements and methods proposed for the execution of the Works. f) Not make significant alteration to the arrangements and methods without previous Notice to the Employer's Representative. g) Attend all meetings as reasonably required by the Employer or Employer's Representative.	a) None. b) None. c) None. d) None. e) Whenever required by the Employer's Representative. f) None. g) None.	None.
4.2 Performance Security	a) Obtain a Performance Security for proper performance and deliver it to the Employer. b) Ensure that the Performance Security is valid and enforceable until the issue of the Contract Completion Certificate. c) Extend the validity of the Performance Security until the Works and the Operation Service have been completed.	a) Within 28 days after receiving the Letter of Acceptance. b) None. c) As required.	Employer may terminate the Contract.

CLAUSE	OBLIGATIONS	TIME FRAME	SPECIFIC CONSEQUENCES OF NON-COMPLIANCE
4.3 Contractor's Representative	a) Appoint the Contractor's Representative and give him all authority necessary to act on the Contractor's behalf under the Contract. b) Submit to the Employer for consent, the name and particulars of the person the Contractor proposes to appoint. c) If consent is withheld or subsequently revoked, or if the appointed person fails to act, submit the name and particulars of another suitable person for such appointment. d) Not to, without the prior consent of the Employer's Representative, revoke the appointment of the Contractor's Representative or appoint a replacement.	a) None. b) Prior to the Commencement Date. c) None. d) None.	None.
4.4 Subcontractors	a) Not subcontract the whole of the Works. b) Be responsible for the acts or defaults of any Subcontractor, his agents or employees. c) Obtain consent of the Employer's Representative to proposed Subcontractors. d) Give the Employer's Representative, Notice of the intended date of the commencement of each Subcontractor's work and the commencement of such work on the Site.	a) None. b) None. c) Prior to appointment. d) Not less than 28 days.	None.
4.6 Co-operation	a) Allow appropriate opportunities for carrying out work to the Employer's Personnel, any other contractors employed by the Employer and the personnel of any legally constituted public authorities. b) Be responsible for the Contractor's construction and operation activities and co-ordinate activities with those of other contractors to the extent specified in the Employer's Requirements. c) Submit such documents which require the Employer to give to the Contractor possession of any foundation, structure, plant or means of access.	a) None. b) None. c) In the time and manner stated in the Employer's Requirements.	None.

THE OBLIGATIONS OF THE CONTRACTOR

Gold Book

THE OBLIGATIONS OF THE CONTRACTOR (continued)

CLAUSE	OBLIGATIONS	TIME FRAME	SPECIFIC CONSEQUENCES OF NON-COMPLIANCE
4.7 Setting Out	a) Set out the Works in relation to original points, lines and levels of reference specified in the Contract or notified by the Employer's Representative. b) Be responsible for the correct positioning of all parts of the Works. c) If the Contractor suffers delay and/or incurs Costs from executing work which was necessitated by an error in the items of reference, give Notice to the Employer's Representative.	a) None. b) None. c) As soon as practicable and not later than 28 days after the Contractor became aware, or should have become aware of the event or circumstance (Sub-Clause 20.1).	a) None. b) None. c) Loss of entitlement to an extension to the Time for Completion and additional payment (Sub-Clause 20.1).
4.8 Safety Procedures	a) Comply with all applicable safety regulations. b) Take care for the safety of all persons entitled to be on the Site. c) Use reasonable efforts to keep the Site and Works clear of unnecessary obstruction. d) Provide fencing, lighting, guarding and watching of the Works. e) Provide any Temporary Works, which may be necessary for the use and protection of the public and of owners and occupiers of adjacent land.	None.	None.
4.9 Quality Assurance	Institute a quality assurance system and submit details to the Employer's Representative.	Before each design, execution and operation stage is commenced.	None.
4.10 Site Data	Be responsible for verifying and interpreting data provided by the Employer on subsurface and hydrological conditions at the Site, including environmental aspects.	None.	None.

CLAUSE	OBLIGATIONS	TIME FRAME	SPECIFIC CONSEQUENCES OF NON-COMPLIANCE
4.12 Unforeseeable Physical Conditions	a) If the Contractor encounters adverse physical conditions which the Contractor considers to have been Unforeseeable, give Notice to the Employer's Representative. b) Continue executing the Works using appropriate measures. c) Comply with any instructions from the Employer's Representative. d) If the Contractor suffers delay and/or incurs Costs, give Notice to the Employer's Representative.	a) As soon as practicable. b) None. c) None. d) As soon as practicable and not later than 28 days after the Contractor became aware, or should have become aware of the event or circumstance (Sub-Clause 20.1).	a) None. b) None. c) None. d) Loss of entitlement to an extension to the Time for Completion and additional payment (Sub-Clause 20.1).
4.13 Rights of Way and Facilities	a) Bear all costs and charges for special and/or temporary rights of way. b) Obtain any additional facilities outside the Site which the Contractor may require for the purposes of the Works.	None.	None.
4.14 Avoidance of Interference	Not to interfere with the convenience of the public, or the access to and use and occupation of all roads and footpaths.	None.	None.
4.15 Access Route	a) Use reasonable efforts to prevent any road or bridge from being damaged. b) Be responsible for any maintenance which may be required for the use of access routes. c) Provide all necessary signs or directions along access routes. d) Obtain any permission which may be required from the relevant authorities for use of routes, signs and directions.	None.	None.
4.16 Transport of Goods	a) Give the Employer's Representative, Notice of the date on which any Plant or a major item of other Goods will be delivered to the Site. b) Be responsible for packing, loading, transporting, receiving, unloading, storing and protecting all Goods and other things required for the Works or Operation Service.	a) 21 days before delivery. b) None.	None.

THE OBLIGATIONS OF THE CONTRACTOR

Gold Book

THE OBLIGATIONS OF THE CONTRACTOR (continued)

CLAUSE	OBLIGATIONS	TIME FRAME	SPECIFIC CONSEQUENCES OF NON-COMPLIANCE
4.17 Contractor's Equipment	a) Be responsible for all Contractors' Equipment. b) Not remove from Site any major items of Contractor's Equipment without the consent of the Employer's Representative.	None.	None.
4.18 Protection of the Environment	a) Take all reasonable steps to protect the environment and to limit damage and nuisance to people and property. b) Ensure that emissions, surface discharges and effluent shall not exceed the values indicated in the Employer's Requirements and shall not exceed the values prescribed by applicable Laws.	None.	None.
4.19 Electricity, Water and Gas	a) Be responsible for the provision of all power, water and other services. b) In a case that supplies of electricity, water, gas and other services are available on the Site and are detailed in the Employer's Requirements, take over the services in the Contractor's name and pay the utility service providers for usage.	None.	None.
4.20 Employer's Equipment and Free-Issue Materials	a) Be responsible for the Employer's Equipment when used by the Contractor. b) Pay the Employer for the use of the Employer's Equipment. c) Inspect free-issue materials. d) Give Notice of any shortage, defect or default in the free-issue materials.	a) None. b) None. c) None. d) Promptly.	None.
4.21 Progress Reports	Prepare and submit monthly progress reports.	Monthly, within 7 days of the period to which the report relates.	None.

CLAUSE	OBLIGATIONS	TIME FRAME	SPECIFIC CONSEQUENCES OF NON-COMPLIANCE
4.22 Security of the Site	a) Be responsible for the security of the Site. b) Keep unauthorised persons off the Site.	None.	None.
4.23 Contractor's Operations on Site	a) Confine operations to the Site and to any additional areas agreed as working areas. b) Take all necessary precautions to keep the Contractor's Equipment and Contractor's Personnel within the Site and any agreed working areas. c) Keep the Site free from all unnecessary obstruction. d) Store or dispose of any Contractor's Equipment or surplus materials. e) Clear away and remove from the Site any wreckage, rubbish and Temporary Works. f) Leave the Site and the Works in a clean and safe condition.	None.	None.
4.24 Fossils	a) Take reasonable precautions to prevent the Contractor's Personnel or other persons from removing or damaging fossils, coins, articles of value or antiquity, structures and other remains or items of geological or archaeological interest. b) Give Notice of the finding of such items. c) Give further Notice if the Contractor suffers delay and/or incurs Cost.	a) None. b) Upon discovery. c) As soon as practicable and not later than 28 days after the Contractor became aware, or should have become aware of the event or circumstance (Sub-Clause 20.1).	a) None. b) None. c) Loss of entitlement to an extension to the Time for Completion and additional payment (Sub-Clause 20.1).
5 Design			
5.1 General Design Obligations	a) Carry out and be responsible for the design of the Works. b) Submit to the Employer's Representative for consent, the names and particulars of each proposed designer and design Subcontractor. c) Scrutinise the Employer's Requirements. d) Give Notice to the Employer's Representative of any error, fault or other defect found in the Employer's Requirements.	a) None. b) None. c) Upon receiving Notice of the Commencement Date. d) Within the period stated in the Contract Data.	None.

THE OBLIGATIONS OF THE CONTRACTOR (continued)

CLAUSE	OBLIGATIONS	TIME FRAME	SPECIFIC CONSEQUENCES OF NON-COMPLIANCE
5.2 Contractor's Documents	a) Prepare all Contractor's Documents. b) Prepare all other documents necessary to instruct the Contractor's Personnel. c) In the case that the Employer's Requirements require it, submit the Contractor's Documents to the Employer's Representative for review. d) Give Notice of the submission of Contractor's Documents to the Employer's Representative for review. e) In the case that the Employer's Representative gives Notice that a Contractor's Document fails to confirm with the Contract, rectify and resubmit the document for review. f) Not to commence execution of such part of the Works until the review period has expired. g) Execute parts of the Works in accordance with the documents submitted for review. h) Give Notice to the Employer's Representative in the case that the Contractor wishes to modify any design or document which has previously been submitted for review. i) Submit revised documents to the Employer's Representative in accordance with the Sub-Clause 5.2 procedure.	a) None. b) None. c) None. d) None. e) None. f) None. g) None. h) Immediately. i) None.	None.
5.3 Contractor's Undertaking	a) In the case that the Employer's Representative instructs that further Contractor's Documents are required, prepare such documents. b) Execute and provide the completed Works in accordance with the Laws and the documents forming the Contract, as altered or modified by Variations.	a) Promptly. b) None.	None.

CLAUSE	OBLIGATIONS	TIME FRAME	SPECIFIC CONSEQUENCES OF NON-COMPLIANCE
5.4 Technical Standards and Regulations	a) Design, provide the Contractor's Documents, execute and provide the completed Works in accordance with the Country's technical standards, building, construction and environmental Laws, Laws applicable to the product being produced from the Works and other standards specified in the Employer's Requirements. b) Give Notice to the Employer's Representative in the case that new or applicable standards come into force after the Base Date and if appropriate, submit proposals for compliance.	None.	None.
5.5 As-Built Documents	a) Prepare and keep up to date a complete set of "as-built" records of the execution of the Works. b) Provide two copies of the as-built records to the Employer's Representative. c) Supply and submit as-built drawings of the Works to the Employer's Representative for review. d) Obtain the consent of the Employer's Representative as to the size, referencing system and other relevant details of the as-built drawings. e) Supply to the Employer's Representative the specified numbers of as-built drawings.	a) None. b) Prior to the commencement of the Tests on Completion of Design-Build. c) None. d) None. e) Prior to the issue of the Commissioning Certificate.	a) None. b) None. c) None. d) None. e) The Works shall not be considered to be completed for the purposes of issuing the Commissioning Certificate.
5.6 Operation and Maintenance Manuals	a) Supply to the Employer's Representative two copies of all operation and maintenance manuals in sufficient detail for the Employer to operate, maintain, dismantle, reassemble, adjust and repair the Plant and the Works. b) Supply the balance of the required operation and maintenance manuals.	a) Prior to the commencement of the Commissioning Period. b) Prior to issue of the Commissioning Certificate.	a) None. b) The Works shall not be considered to be completed for the purposes of issuing the Commissioning Certificate.
5.7 Design Error	Correct any errors, omissions, ambiguities, inconsistencies, inadequacies or other defects that are found in the Contractor's Documents.	None.	None.

Gold Book

THE OBLIGATIONS OF THE CONTRACTOR (continued)

CLAUSE	OBLIGATIONS	TIME FRAME	SPECIFIC CONSEQUENCES OF NON-COMPLIANCE
6 Staff and Labour			
6.1 Engagement of Staff and Labour	Make arrangements for the engagement of all staff and labour, local or otherwise and for their payment, housing, feeding and transport.	None.	None.
6.2 Rates of Wages and Conditions of Labour	Pay rates of wages and observe conditions of labour which are not lower than those established for the trade or industry where the work is carried out.	None.	None.
6.3 Persons in the Service of Employer	Not recruit, or attempt to recruit, staff and labour from amongst the Employer's Personnel.	None.	None.
6.4 Labour Laws	a) Comply with all the relevant labour Laws. b) Require employees to obey all applicable Laws.	None.	None.
6.5 Working Hours	Obtain the consent of the Employer's Representative if working outside the normal working hours.	None.	None.
6.6 Facilities for Staff and Labour	a) Provide and maintain all necessary accommodation and welfare facilities for the Contractor's Personnel. b) Provide facilities for the Employer's Personnel as stated in the Employer's Requirements. c) Not permit any of the Contractor's Personnel to maintain any temporary or permanent living quarters within the Site of the Works unless the Employer has given written permission.	None.	None.

CLAUSE	OBLIGATIONS	TIME FRAME	SPECIFIC CONSEQUENCES OF NON-COMPLIANCE
6.7 Health and Safety	a) Take all reasonable precautions to maintain the health and safety of the Contractor's Personnel. b) Ensure that medical staff, first aid facilities, sick bay and ambulance service are available at all times and that suitable arrangements are made for all necessary welfare and hygiene requirements and for the prevention of epidemics. c) Appoint a qualified accident prevention officer and whatever is required by this person to exercise this responsibility and authority. d) Send to the Employer's Representative details of any accident. e) Maintain records and make reports concerning health, safety, welfare and damage to property.	a) None. b) None. c) None. d) As soon as practicable after its occurrence. e) None.	None.
6.8 Contractor's Superintendence	Provide all necessary superintendence to plan, arrange, direct, manage, inspect, test and monitor the design and execution of the Works and the provision of the Operation Service.	None.	None.
6.9 Contractor's Personnel	In the case that the Employer's Representative requires the removal of any person employed on the Site or Works and if appropriate, appoint a suitable replacement person.	None.	None.
6.10 Records of Contractor's Personnel and Equipment	a) During the Design-Build period, submit to the Employer's Representative, details showing the number of each class of Contractor's Personnel and of each type of Contractor's Equipment on the Site. b) Notify the Employer's Representative of changes to the Equipment. c) During the Operation Service Period, notify the Employer's Representative of any changes to the Personnel or Equipment.	a) None b) At the end of each calendar month. c) At the end of each calendar month.	None.
6.11 Disorderly Conduct	Take all reasonable precautions to prevent any unlawful, riotous or disorderly conduct by, or amongst the Contractor's Personnel and to preserve peace and protection of persons and property on and near the Site.	None.	None.

THE OBLIGATIONS OF THE CONTRACTOR (continued)

CLAUSE	OBLIGATIONS	TIME FRAME	SPECIFIC CONSEQUENCES OF NON-COMPLIANCE
7 Plant, Materials and Workmanship			
7.1 Manner of Execution	Carry out the manufacture and/or replacement and/or repair of Plant, the production and manufacture of Materials and all other activities during the execution of the Works and provision of the Operation Service.	None.	None.
7.2 Samples	Submit samples of Materials and relevant information to the Employer's Representative for review.	None.	None.
7.3 Inspection	a) Give the Employer's Personnel full opportunity to carry out inspections. b) Give Notice to the Employer's Representative to inspect.	a) None. b) Whenever any work is ready and before it is covered up, put out of sight or packaged for storage or transport.	a) None. b) Uncover the work, reinstate and make good at the Contractor's cost.
7.4 Testing	a) Provide everything necessary to carry out the specified tests. b) Agree with the Employer's Representative, the time and place for the testing. c) If the Contractor suffers delay and/or incurs Costs, give Notice to the Employer's Representative. d) Forward to the Employer's Representative, certified reports of the tests.	a) None. b) None. c) As soon as practicable and not later than 28 days after the Contractor became aware, or should have become aware of the event or circumstance (Sub-Clause 20.1). d) Promptly.	a) None. b) None. c) Loss of entitlement to an extension to the Time for Completion and additional payment (Sub-Clause 20.1). d) None.
7.5 Rejection	Make good defects notified by the Employer's Representative.	Promptly.	None.

CLAUSE	OBLIGATIONS	TIME FRAME	SPECIFIC CONSEQUENCES OF NON-COMPLIANCE
7.6 Remedial Work	Comply with the instructions of the Employer's Representative with regard to remedial work.	Within a reasonable time.	Contractor shall pay Costs incurred by the Employer in engaging other persons to carry out the work.
7.8 Royalties	Pay all royalties, rents and other payments for natural Materials obtained from outside the Site and disposal of surplus materials.	None.	None.
8 Commencement Date, Completion and Programme			
8.2 Time for Completion	a) Complete the whole of the Works and each Section within the times specified in the Contract. b) Provide the Operation Service for the period stated in the Contract Data.	None.	a) Contractor shall pay delay damages to the Employer (Sub-Clause 8.5). b) Contractor shall pay compensation to the Employer (Sub-Clause 8.5).
8.3 Programme	a) Submit a detailed time programme. b) Submit a revised programme. c) Proceed in accordance with the programme. d) Submit a revised programme on receiving a Notice from the Employer's Representative that a programme fails to comply with the Contract, or to be consistent with actual progress.	a) Within 28 days of receiving Notice of the Commencement Date. b) Whenever the previous programme is inconsistent with actual progress or with the Contractor's obligations. c) None. d) Within 14 days of the Notice.	None.

THE OBLIGATIONS OF THE CONTRACTOR (continued)

CLAUSE	OBLIGATIONS	TIME FRAME	SPECIFIC CONSEQUENCES OF NON-COMPLIANCE
8.4 Advance Warning	Advise the other Party in advance of any known or probable future events or circumstances which may adversely affect the work, increase the Contract Price or delay the execution of the Works or the Operation Service.	None.	None.
8.5 Delay Damages	a) Pay delay damages in the case of failure to comply with the Time for Completion of Design-Build. b) Pay compensation in the case that the Contractor fails or is unable to provide the Operation Service.	None.	None.
8.7 Handback Requirements	Ensure that the Works comply with the handback requirements specified in the Employer's Requirements.	Prior to the issue of the Contract Completion Certificate.	None.
9 Design-Build			
9.1 Commencement of Design-Build	Commence the design and execution of the Works and proceed with the Design-Build with due expedition and without delay.	Within 28 days of the Commencement Date.	None.
9.2 Time for Completion of Design-Build	Complete the whole of the Design-Build of the Works and each Section.	Within the Time for Completion of Design-Build of the Works or Section as set out in the Contract Data.	Contractor shall pay delay damages to the employer (Sub-Clause 9.6).
9.3 Extension of Time for Completion of Design-Build	Give Notice to the Employer's Representative if the Contractor considers himself to be entitled to an extension of the Time for Completion.	As soon as practicable and not later than 28 days after the Contractor became aware, or should have become aware of the event or circumstance (Sub-Clause 20.1).	Loss of entitlement to an extension to the Time for Completion (Sub-Clause 20.1).

CLAUSE	OBLIGATIONS	TIME FRAME	SPECIFIC CONSEQUENCES OF NON-COMPLIANCE
9.5 Rate of Progress	Adopt revised methods in order to expedite progress and complete within the Time for Completion of Design-Build.	None.	None.
9.7 Suspension of Work	Protect, store and secure such part or the Works in the case of an instruction to suspend the Works.	None.	None.
9.8 Consequences of Suspension	Give Notice to the Employer's Representative if the Contractor suffers delay and/or incurs Cost from complying with the Employer's Representative's instructions under this clause or from resuming work.	As soon as practicable and not later than 28 days after the Contractor became aware, or should have become aware of the event or circumstance (Sub-Clause 20.1).	Loss of entitlement to an extension to the Time for Completion and additional payment (Sub-Clause 20.1).
9.11 Resumption of Work	a) Jointly examine the Works and the Plant and Materials affected by the suspension with the Employer's Representative. b) Make good any deterioration, defect or loss.	None.	None.
10 Operation Service			
10.1 General Requirements	a) Comply with the Operation Management Requirements and any agreed revisions thereof. b) Follow the requirement for the Operation and Maintenance Plan and manuals. c) Ensure that the Works remain fit for purpose.	None.	None.
10.2 Commencement of Operation Service	a) Comply with requirements and/or restrictions contained in the Commissioning Certificate or any relevant Notice. b) Provide the Operation Service in compliance with the Operation Management Requirements. c) In the case that the Contractor wishes to modify an approved document, notify the Employer's Representative. d) Not to implement any modification until consent has been given by the Employer's Representative.	a) None. b) From the date stated in the Commissioning Certificate. c) Immediately. d) None.	None.

THE OBLIGATIONS OF THE CONTRACTOR

Gold Book

THE OBLIGATIONS OF THE CONTRACTOR (continued)

CLAUSE	OBLIGATIONS	TIME FRAME	SPECIFIC CONSEQUENCES OF NON-COMPLIANCE
10.3 Independent Compliance Audit	a) Jointly with the Employer, appoint the Auditing Body to carry out an independent audit during the Operation Service. b) Cooperate with the Auditing Body and give due regard to each report issued by the Auditing Body.	a) At least 182 days prior to the commencement of the Operation Service. b) None.	a) The DAB shall make the appointment. b) None.
10.5 Training	Carry out the training of the Employer's Personnel.	To be agreed with the Employer.	None.
10.6 Delays and Interruptions during the Operation Service	a) In the case of the Employer's Representative issuing an instruction to suspend progress of the Operation Service, protect, store, secure and maintain the Plant against any deterioration, loss or damage. b) Jointly examine the Works with the Employer's Representative after the period of suspension. c) Make good any deterioration or defect in the Plant.	None.	None.
10.7 Failure to Reach Production Outputs	a) In the event that the Contractor fails to achieve production outputs required under the Contract, establish the cause of such failure jointly with the Employer. b) If the cause of the failure lies with the Contractor, take all steps necessary to restore the output.	None.	None.

CLAUSE	OBLIGATIONS	TIME FRAME	SPECIFIC CONSEQUENCES OF NON-COMPLIANCE
11 Testing			
11.1 Testing of the Works	a) Carry out the Tests on Completion of Design-Build. b) Give Notice to the Employer's Representative of the date after which the Contractor will be ready to carry out each of the Tests on Completion of Design-Build. c) In the case of trial operation, give the Employer's Representative Notice that the Works are ready for any other Tests on Completion of Design-Build. d) Submit a certified report of the results of the Tests to the Employer's Representative.	a) After providing the documents in accordance with Sub-Clauses 5.5 and 5.6. b) Not less than 21 days. c) When the Works are operating under stable conditions. d) As soon as the Works, or a Section have passed the Tests on Completion of Design-Build.	None.
11.2 Delayed Tests on Completion of Design-Build	Carry out the Tests if the Employer's Representative gives Notice of undue delay.	Within 21 days of the Employer's Representative Notice.	The Employers personnel may proceed with the tests at the Contractor's Cost.
11.5 Completion of the Works and Sections	In the case that the Employer's Representative specifies work to be done by the Contactor to enable the Commissioning Certificate to be issued, complete the specified work.	None.	The Commissioning Certificate will not be issued.
11.8 Joint Inspection Prior to Contract Completion	a) Jointly, with the Employer's Representative, carry out an inspection of the Works. b) Submit a report on the condition of the Works. c) Submit a programme for carrying out any maintenance, replacement or other works identified in the report.	a) Not less than two years prior to the expiry date of the Operation Service Period. b) Within 28 days of the completion of the joint inspection. c) None.	None.

THE OBLIGATIONS OF THE CONTRACTOR

Gold Book

THE OBLIGATIONS OF THE CONTRACTOR (continued)

CLAUSE	OBLIGATIONS	TIME FRAME	SPECIFIC CONSEQUENCES OF NON-COMPLIANCE
11.9 Procedure for Tests Prior to Contract Completion	a) Carry out the Tests Prior to Completion. b) Provide all necessary labour, materials, electricity, fuel and water. c) Undertake any required remedial works. d) Jointly, with the Employer's Representative, compile and evaluate the results of the Tests. e) Make the results of any tests, inspections or monitoring available to the Employer's Representative. f) Notify the Employer's Representative that the Works are complete and ready for final inspection.	a) Towards the end of the Operation Service Period. b) As necessary. c) None. d) None. e) None. f) As soon as the Contractor has completed the Tests.	None.
12 Defects			
12.1 Completion of Outstanding Work and Remedying Defects	a) Complete any work which is outstanding on the date stated in a Commissioning Certificate. b) Execute all work required to remedy defects or damage. c) Repair or make good any damage or defect occurring during the Operation Service Period.	a) As soon as practicable and not later than one year after the date stated in the Commissioning Certificate. b) None. c) None.	a&b) Final payment for the Design-Build Period will not be certified. c) The Contract Completion Certificate will not be issued.
12.2 Cost of Remedying Defects	Where the Contractor is required to remedy a defect or damage that is attributable to the Employer, notify the Employer's Representative.	None.	None.
12.6 Contractor to Search	If required by the Employer's Representative, search for the cause of any defect.	None.	None.

CLAUSE	OBLIGATIONS	TIME FRAME	SPECIFIC CONSEQUENCES OF NON-COMPLIANCE
13 Variations and Adjustments			
13.1 Right to Vary	a) Execute and be bound by each Variation. b) In the case that the Employer or the Employer's Representative instructs a Variation during the Operation Service Period, proceed in accordance with Sub-*Clause 13.3 (Variation Procedure)*.	None.	None.
13.3 Variation Procedure	a) Respond in writing to a request for a proposal. b) Not delay any work whilst awaiting a response. c) Acknowledge receipt of Variation instructions.	a) As soon as practicable. b) None. c) None.	None.
13.5 Provisional Sums	Produce quotations, invoices, vouchers and accounts or receipts in substantiation of the amounts paid to nominated Subcontractors.	When required by the Employer's Representative.	None.
13.6 Adjustments for Changes in Legislation	If the Contractor suffers (or will suffer) delay and/or incurs (or will occur) additional Cost, give Notice to the Employer's Representative.	As soon as practicable and not later than 28 days after the Contractor became aware, or should have become aware of the event or circumstance (Sub-Clause 20.1).	Loss of entitlement to an extension to the Time for Completion and additional payment (Sub-Clause 20.1).
14 Contract Price and Payment			
14. 1 The Contract Price	Pay all taxes, duties and fees required to be paid under the Contract.	None.	None.
14.2 Advance Payment	a) Submit an advance payment guarantee. b) Extend the validity of the guarantee until the advance payment has been repaid.	None.	a) Employer is not obligated to make the advance payment. b) None.

THE OBLIGATIONS OF THE CONTRACTOR (continued)

CLAUSE	OBLIGATIONS	TIME FRAME	SPECIFIC CONSEQUENCES OF NON-COMPLIANCE
14.3 Application for Advance and Interim Payment Certificates	a) Submit an application for the advance payment. b) Submit, one original and five copies of a Statement showing in detail the amounts to which the Contractor considers himself to be entitled.	a) When submitting the advance payment guarantee. b) After the end of each month.	No obligation on the Employer's Representative to certify payment (Sub-Clause 14.7).
14.4 Schedule of Payments	In the case that the Contract does not include a schedule of payments, submit non-binding estimates of the payments expected to become due.	a) First estimate within 42 days after the Commencement Date. b) Revised estimates at quarterly intervals.	None.
14.11 Application for Final Payment Certificate Design-Build	a) Submit one original and five copies of a Final Statement Design-Build with supporting documents. b) Submit a written undertaking that the Statement is in full and final settlement relating to the Design-Build. c) In the case that the Employer's Representative disagrees with the Final Statement, attempt to agree such matters. d) Re-submit the Final Statement as agreed with the Employer's Representative.	a) Within 28 days after the end of the Retention Period. b) Together with the Final Statement. c) None. d) None	a) Employer's Representative will issue an Interim Payment Certificate for the amount that he considers due. b) None. c) Employer's Representative will issue an Interim Payment Certificate for the amount that he considers due. d) None.

CLAUSE	OBLIGATIONS	TIME FRAME	SPECIFIC CONSEQUENCES OF NON-COMPLIANCE
14.13 Application for Final Payment Certificate Operation Service	a) Submit one original and five copies of a Final Statement Operation Service with supporting documents. b) Submit a written discharge according to the requirements of Sub-Clause 14.14 *(Discharge)*.	a) Within 56 days after receiving the Contract Completion Certificate. b) Together with the Final Statement.	The Employer's Representative is not obligated to issue a Final Payment Certificate Operation Service.
14.14 Discharge	Submit a written discharge.	When submitting the Final Statement Operation Service.	None.
14.15 Issue of Final Payment Certificate Operation Service	In the case that the Employer's Representative disagrees with the Final Statement, attempt to agree such matters.	None.	Employer's Representative will issue a Final Payment Certificate for the amount that he considers due.
14.18 Asset Replacement Fund	Give Notice to the Employer's Representative of intention to replace any item of Plant identified in the Asset Replacement Schedule.	At least 28 days prior to the intended date.	None.
14.19 Maintenance Retention Fund	Ensure that the Maintenance Retention Guarantee remains valid and in force until the issue of the Contract Completion Certificate.	None.	None.
15 Termination by Employer			
15.2 Termination for Contractor's Default	a) In the case of a Notice of termination being served, leave the Site and deliver any required Goods, all Contractor's Documents and other design documents to the Employer's Representative. b) Use best efforts to comply with any reasonable instructions included in the Notice. c) Arrange for the removal of Equipment and Temporary Works.	a) None. b) Immediately. c) Promptly.	a) None. b) None. c) Items may be sold by the Employer.

THE OBLIGATIONS OF THE CONTRACTOR

Gold Book

THE OBLIGATIONS OF THE CONTRACTOR (continued)

CLAUSE	OBLIGATIONS	TIME FRAME	SPECIFIC CONSEQUENCES OF NON-COMPLIANCE
15.7 Payment after Termination for Employer's Convenience	In the case of a Notice of termination, cease all further work, hand over Contractor's Documents, Plant, Materials and other work and remove all other Goods from the Site.	After termination.	None.
16 Suspension and Termination by Contractor			
16.1 Contractor's Entitlement to Suspend Work	a) Give Notice if the Contractor intends to suspend work or reduce the rate of work. b) Resume normal working when the Employer's obligations have been met and/or the Interim Payment Certificate has been received. c) If the Contractor suffers delay and/or incurs Cost, give Notice to the Employer's Representative.	a) 21 days before the intended suspension or reduction in the rate of work. b) As soon as is reasonably practicable. c) As soon as practicable and not later than 28 days after the Contractor became aware, or should have become aware of the event or circumstance (Sub-Clause 20.1).	a) None. b) None. c) Loss of entitlement to an extension to the Time for Completion and additional payment (Sub-Clause 20.1).
16.2 Termination by Contractor	Give Notice of intention to terminate.	14 days before the intended termination date.	None.
16.3 Cessation of Work and Removal of Contractor's Equipment	In the case of a Notice of termination, cease all further work, hand over Contractor's Documents, Plant, Materials and other work and remove all other Goods from the Site.	After the Notice has taken effect.	None.

CLAUSE	OBLIGATIONS	TIME FRAME	SPECIFIC CONSEQUENCES OF NON-COMPLIANCE
17 Risk Allocation			
17.5 Responsibility for Care of the Works	a) Take full responsibility for the care of the Works and Goods. b) Be responsible for the care of the Permanent Works. c) Be responsible for the care of any part of the Permanent Works for which a Section Commissioning Certificate has been issued. d) Take responsibility for any outstanding work which is to be completed during the Operation Service Period.	a) From the Commencement Date until the Commissioning Certificate for the whole of the Works is issued. b) During the Operation Service Period. c) None. d) Until the outstanding work is completed.	None.
17.6 Consequences of the Employer's Risks of Damage	a) Give Notice in the case of an Employer's risk event that results in loss or damage. b) If the Contractor suffers delay and/or incurs Cost from rectifying loss or damage, give further Notice to the Employer's Representative.	a) Promptly. b) As soon as practicable and not later than 28 days after the Contractor became aware, or should have become aware of the event or circumstance (Sub-Clause 20.1).	a) None. b) Loss of entitlement to an extension to the Time for Completion and additional payment (Sub-Clause 20.1).
17.7 Consequences of the Contractor's Risks resulting in Damage	a) Give Notice in the case of a Contractor's risk event that results in loss or damage. b) Rectify the damage.	a) Promptly. b) None.	None.

THE OBLIGATIONS OF THE CONTRACTOR

Gold Book

THE OBLIGATIONS OF THE CONTRACTOR (continued)

CLAUSE	OBLIGATIONS	TIME FRAME	SPECIFIC CONSEQUENCES OF NON-COMPLIANCE
17.9 Indemnities by the Contractor	a) Indemnify and hold harmless the Employer, the Employer's Personnel and their respective agents against and from all claims, damages, losses and expenses in respect of bodily injury, sickness, disease, death, damage to or of loss of property by reason of the Contractor's design, the execution and completion or operation and maintenance of the Works. b) Indemnify the Employer against all errors in the Contractor's design of the Works and other professional services which result in the Works not being fit for purpose or result in loss and/or damage for the Employer.	None.	None.
17.12 Risk of Infringement of Intellectual and Industrial Property Rights	a) Indemnify and hold the Employer harmless against and from any other claim which arises out of, or in relation to the Contractor's design, manufacture, construction or execution of the Works, the use of Contractor's Equipment or the proper use of the Works. b) If requested by the Employer, assist in contesting the claim. c) Not make any admission which might be prejudicial to the Employer.	a) None. b) None. c) None.	a) None. b) None. c) None.
18 Exceptional Risks			
18.3 Duty to Minimise Delay	a) Use all reasonable endeavours to minimise any delay. b) Give Notice to the Employer when the effects of the Exceptional Event cease.	None.	None.

CLAUSE	OBLIGATIONS	TIME FRAME	SPECIFIC CONSEQUENCES OF NON-COMPLIANCE
19 Insurance			
19.1 General Requirements	a) Effect and maintain the insurances for which the Contractor is responsible with insurers and in terms consistent with any terms to be agreed with the Employer. b) If required, produce the insurance policies. c) Send copies of the receipts of payments of the premiums to the Employer. d) Notify the insurers of any relevant changes of the Works. e) Notify the insurers of any changes in the Operation Service. f) Ensure that the insurance is maintained adequately during the performance of the Contract.	a) None. b) When required by the Employer. c) Upon payment of premium. d) As appropriate. e) As appropriate. f) None.	Employer may effect the insurances and recover the cost from the Contractor.
19.2 Insurances to be provided by the Contractor during the Design-Build Period	a) Provide insurance for the Works, in the joint names of the Contractor and the Employer. b) Provide insurance for Contractor's Equipment, in the joint names of the Contractor and the Employer. c) Provide insurance against legal liability arising out of negligent fault, defect, error or omission of the Contractor. d) Provide insurance against liabilities for death or injury to any person, or loss of or damage to property, in the joint names of the Contractor and the Employer. e) Provide insurance against liability for claims, damages, losses and expenses arising from injury, sickness, disease or death. f) Provide any other insurance required by Law and local practice as detailed in the Contract Data.	During the Design-Build Period.	Employer may effect the insurances and recover the cost from the Contractor (Sub-Clause 19.1).

THE OBLIGATIONS OF THE CONTRACTOR (continued)

CLAUSE	OBLIGATIONS	TIME FRAME	SPECIFIC CONSEQUENCES OF NON-COMPLIANCE
19.3 Insurances to be provided by the Contractor during the Operation Service Period	a) Provide fire extended cover insurance for the Works, in the joint names of the Contractor and the Employer. b) Submit the terms of the fire extended cover insurance policy for the Employer's approval. c) Provide insurance against liabilities for death or injury to any person, or loss of or damage to property, in the joint names of the Contractor and the Employer. d) Provide insurance against liability for claims, damages, losses and expenses arising from injury, sickness, disease or death. e) Provide any other insurance required by Law and local practice as detailed in the Contract Data. f) Provide any other operational insurances detailed in the Contract Data.	a) From the date stated in the Commissioning Certificate. b) No later than 28 days before the Commissioning Certificate is due to be issued. c) Prior to the issue of the Commissioning Certificate until the issue of the Contract Completion Certificate. d) Prior to the issue of the Commissioning Certificate until the issue of the Contract Completion Certificate or when the last of his employees and subcontractors have left the Site. e) As detailed in the Contract Data. f) As detailed in the Contract Data.	Employer may effect the insurances and recover the cost from the Contractor (Sub-Clause 19.1).

CLAUSE	OBLIGATIONS	TIME FRAME	SPECIFIC CONSEQUENCES OF NON-COMPLIANCE
20 Claims, Disputes and Arbitration			
20.1 Contractor's Claims	a) Give Notice if the Contractor considers himself to be entitled to any extension of the Time for Completion of Design-Build and/or any additional payment. b) Keep such contemporary records as may be necessary to substantiate any claim and permit the Employer's Representative to inspect all the records. c) Send to the Employer's Representative a fully detailed claim. d) Provide the Employer's Representative with any additional particulars that may be reasonably required. e) Send further interim claims if the event or circumstance giving rise to the claim has a continuing effect. f) Send a final claim.	a) As soon as practicable and not later than 28 days after the Contractor became aware, or should have become aware of the event or circumstance. b) None. c) Within 42 days after the Contractor became aware (or should have become aware) of the event or circumstance giving rise to the claim. d) None. e) At 28 day intervals. f) Within 28 days after the end of the effects resulting from the event or circumstance.	a) Loss of entitlement to an extension to the Time for Completion and additional payment. b) None. c) The Notice will be deemed to have lapsed and will no longer be valid. d) None. e) None. f) None.
20.3 Appointment of the Dispute Adjudication Board	a) Jointly with the Employer, appoint the Dispute Adjudication Board (DAB). b) Mutually agree the terms of remuneration for the DAB. c) Not to act alone in the termination of any member of the DAB.	a) By the date stated in the Contract Data. b) None. c) None.	a) The appointing entity or official named in the Contract Data shall make the appointment (Sub-Clause 20.4). b) None. c) None.

THE OBLIGATIONS OF THE CONTRACTOR

Gold Book

THE OBLIGATIONS OF THE CONTRACTOR (continued)

CLAUSE	OBLIGATIONS	TIME FRAME	SPECIFIC CONSEQUENCES OF NON-COMPLIANCE
20.6 Obtaining Dispute Adjudication Board's Decision	a) Make available to the DAB, all information, access to the Site and appropriate facilities as the DAB may require. b) Comply with a DAB decision. c) Continue to proceed with the Works in accordance with the Contract.	a) Promptly. b) Promptly. c) None.	a) None. b&c) The matter may be referred to arbitration (Sub-Clause 20.8).
20.7 Amicable Settlement	In the case of a Notice of dissatisfaction being issued, attempt to settle the dispute amicably.	Before the commencement of arbitration.	Arbitration may be commenced.
20.10 Disputes Arising during the Operation Service Period	a) Jointly with the Employer, appoint the one-person DAB. b) Appoint a new Operation Service DAB.	a) At the time of issue of the Commissioning Certificate. b) At the end of each five-year period.	The appointing entity or official named in the Contract Data shall make the appointment (Sub-Clause 20.4).
GENERAL CONDITIONS OF DISPUTE ADJUDICATION AGREEMENT			
5 General Obligations of the Employer and the Contractor	a) Not to request advice from, or consult with the Member regarding the Contract, otherwise than in the normal course of the DAB's activities or when both Parties jointly agree to refer a matter to the DAB. b) In the case that the Contractor refers a Dispute to the DAB which will require the DAB to make a Site visit and attend a hearing, provide appropriate security for a sum equivalent to the DAB's expenses.	None.	None.

CLAUSE	OBLIGATIONS	TIME FRAME	SPECIFIC CONSEQUENCES OF NON-COMPLIANCE
6 Payment	a) Pay each of the DAB member's invoices in full. b) Apply to the Employer for reimbursement of one-half of the DAB invoices by way of the Statements.	a) Within 56 days of receipt. b) None.	a) Employer entitled to pay fees and recover reimbursement of fees, plus financing charges from the Contractor and the DAB Member may suspend services or resign the appointment. b) None.
Procedural Rules for Dispute Adjudication Board Members			
2.	Agree the timing and agenda for each Site visit with the Employer and the DAB.	None.	Timing and agenda shall be decided by the DAB.
3.	a) Attend Site visits by the DAB. b) Cooperate with the Employer in coordinating the DAB Site visits.	None.	None.
4.	a) Jointly with the Employer, furnish to each member of the DAB one copy of all documents which the DAB may request. b) Copy the Employer on all communications between the DAB and the Contractor.	None.	None.

Gold Book

THE OBLIGATIONS OF THE EMPLOYER'S REPRESENTATIVE

CLAUSE	OBLIGATIONS	TIME FRAME	SPECIFIC CONSEQUENCES OF NON-COMPLIANCE
GENERAL CONDITIONS			
General Provisions			
1.3 Notices and Other Communications	Not to unreasonably withhold or delay Notices and other communications.	None.	None.
1.5 Priority of Documents	In the case that an ambiguity or discrepancy is found in the Contract documents, issue any necessary clarification or instruction.	None.	None.
1.10 Errors in the Employer's Requirements	In the case of a Notice of an error in the Employer's Requirements being received, confirm to the Contractor the information listed in this Sub-Clause.	Promptly.	None.
2 THE EMPLOYER			
2.1 Right of Access to the Site	In the case of a Notice and claim for delay or Cost being received, respond to the claim and agree or determine the matters.	Within 42 days after receiving a claim or any further particulars (Sub-Clause 20.1).	Either Party may refer the matter to the DAB (Sub-Clause 20.1).
3 The Employer's Representative			
3.1 Employer's Representative's Duties and Authority	a) Carry out the duties assigned in the Contract. b) Provide staff that shall include suitably qualified engineers and other professionals who are competent to carry out these duties.	None.	None.

CLAUSE	OBLIGATIONS	TIME FRAME	SPECIFIC CONSEQUENCES OF NON-COMPLIANCE
3.2 Delegation by the Employer's Representative	a) In the case of delegation of the Employer's Representative authority, delegate such authority in writing. b) Not to delegate the authority to determine any matter in accordance with Sub-Clause 3.5 *[Determinations]*. c) In the case that the Contractor questions any determination or instruction of an assistant and refers the matter, confirm, reverse or vary the determination or instruction.	a) None. b) None. c) Promptly.	None.
3.3 Instructions of the Employer's Representative	a) Give instructions in writing. b) In the case that the Contractor considers that any instruction does not comply with applicable Laws or is technically impossible, either confirm or amend the instruction.	None.	None.
3.5 Determinations	a) Consult with each Party in an endeavour to reach agreement. b) If agreement is not achieved, make a fair determination in accordance with the Contract, taking due regard of all relevant circumstances. c) Give Notice to both Parties of each agreement or determination, with supporting particulars.	None.	None.
4 The Contractor			
4.3 Contractor's Representative	Respond to the Contractor's request for consent to the appointment of the Contractor's Representative.	None.	None.
4.4 Subcontractors	Respond to the Contractor's requests for consent for proposed Subcontractors.	None.	None.
4.7 Setting Out	In the case of a Notice and claim for delay or Cost being received, respond to the claim and agree or determine the matters.	Within 42 days after receiving a claim or any further particulars (Sub-Clause 20.1).	Either Party may refer the matter to the DAB (Sub-Clause 20.1).

THE OBLIGATIONS OF THE EMPLOYER'S REPRESENTATIVE (continued)

CLAUSE	OBLIGATIONS	TIME FRAME	SPECIFIC CONSEQUENCES OF NON-COMPLIANCE
4.12 Unforeseeable Physical Conditions	a) In the case of a Notice of unforeseen physical conditions, inspect the physical conditions. b) In the case of a Notice and claim for delay or cost being received, respond to the claim and agree or determine the matters.	a) None. b) Within 42 days after receiving a claim or any further particulars (Sub-Clause 20.1).	a) None. b) Either Party may refer the matter to the DAB (Sub-Clause 20.1).
4.21 Progress Reports	Agree the format of monthly progress reports with the Contractor.	None.	None.
4.23 Contractor's Operations on Site	Cooperate with the Contractor to agree additional working areas outside the Site.	None.	None.
4.24 Fossils	a) Give instructions for dealing with fossils, coins, articles of value or antiquity, and structures and other remains or items of geological or archaeological interest found on the Site. b) In the case of a Notice and claim for delay or cost being received, respond to the claim and agree or determine the matters.	a) None. b) Within 42 days after receiving a claim or any further particulars (Sub-Clause 20.1).	Either Party may refer the matter to the DAB (Sub-Clause 20.1).
5 Design			
5.1 General Design Obligations	a) Respond to the Contractor's requests for consent for proposed design Subcontractors. b) In the case of a Notice of any error, fault or other defect found in the Employer's Requirements, determine whether Clause 13 (Variations and Adjustments) shall be applied and give Notice to the Contractor.	None.	None.

CLAUSE	OBLIGATIONS	TIME FRAME	SPECIFIC CONSEQUENCES OF NON-COMPLIANCE
5.2 Contractor's Documents	a) Review and/or approve the Contractor's Documents as described in the Employer's Requirements. b) Give Notice to the Contractor of approval other otherwise.	Within 21 days of receipt of the Contractor's Documents and Notice, unless otherwise stated in the Employer's Requirements.	The Employer's Representative shall be deemed to have approved the Contractor's Documents and the Contractor may proceed.
5.4 Technical Standards and Regulations	In the case that the Contractor submits a Notice that changed or new applicable standards have come into force, determine whether compliance is required and whether the proposals for compliance constitute a Variation.	None.	None.
5.5 As-Built Documents	a) Review the Contractor's as-built drawings. b) Cooperate with the Contractor to agree the size, the referencing system and other relevant details of the as-built drawings and provide consent.	None.	None.
7 Plant, Materials and Workmanship			
7.2 Samples	Review samples submitted by the Contractor in accordance with the procedures described in Sub-Clause 5.2 (*Contractor's Documents*).	Within 21 days of receipt of the Contractor's Documents and Notice, unless otherwise stated in the Employer's Requirements (Sub-Clause 5.2).	The Employer's Representative shall be deemed to have approved the samples and the Contractor may proceed (Sub-Clause 5.2).
7.3 Inspection	Examine, inspect, measure and test the Materials and workmanship.	Without unreasonable delay (or promptly give Notice that inspection is not required).	None.

THE OBLIGATIONS OF THE EMPLOYER'S REPRESENTATIVE

Gold Book

THE OBLIGATIONS OF THE EMPLOYER'S REPRESENTATIVE (continued)

CLAUSE	OBLIGATIONS	TIME FRAME	SPECIFIC CONSEQUENCES OF NON-COMPLIANCE
7.4 Testing	a) Agree with the Contractor, the time and place for the specified testing. b) Give the Contractor Notice of intention to attend the tests. c) In the case of a Notice and claim for cost being received, respond to the claim and agree or determine the matters. d) Endorse the Contractor's test certificate, or issue a certificate confirming that the tests have been passed.	a) None. b) Not less than 24 hours. c) Respond within 42 days after receiving a claim or any further particulars (Sub-Clause 20.1). d) Promptly.	a) None. b) If the Employer's Representative does not attend, the Contractor may proceed and the tests shall be deemed to have been made in the Employer's Representative's presence. c) Either Party may refer the matter to the DAB (Sub-Clause 20.1). d) None.
7.5 Rejection	In the case of rejection of any Plant, Materials or workmanship as a result of being found to be defective or not in accordance with the Contract, give Notice of rejection.	None.	None.
8 Commencement Date, Completion and Programme			
8.1 Commencement Date	Give the Contractor Notice of the Commencement Date.	Not less than 14 days prior to the Commencement Date and within 42 days after the Contractor receives the Letter of Acceptance.	None.

CLAUSE	OBLIGATIONS	TIME FRAME	SPECIFIC CONSEQUENCES OF NON-COMPLIANCE
8.3 Programme	In the case that a programme does not comply with the Contract, give Notice to the Contractor.	Within 21 days after receiving the programme.	Contractor shall proceed in accordance with the programme.
8.6 Contract Completion Certificate	Issue the Contract Completion Certificate to the Contractor.	Within 21 days after the last day of the Contract Period.	None.
9 Design-Build			
9.3 Extension of Time for Completion of Design-Build	In the case of a Notice and claim for delay being received, respond to the claim and agree or determine the matters.	Within 42 days after receiving a claim or any further particulars (Sub-Clause 20.1).	Either Party may refer the matter to the DAB (Sub-Clause 20.1).
9.7 Suspension of Work	In the case of an instruction to suspend progress, notify the cause for the suspension.	None.	None.
9.8 Consequences of Suspension	In the case of a Notice and claim for delay or cost being received, respond to the claim and agree or determine the matters.	Within 42 days after receiving a claim or any further particulars (Sub-Clause 20.1).	Either Party may refer the matter to the DAB (Sub-Clause 20.1).
9.10 Prolonged Suspension	In the case that the Contractor requests permission to proceed after 84 days of suspension, respond to the Contractor's request.	Within 28 days of the request.	Contractor may treat the suspension as an omission or give Notice of termination.
9.11 Resumption of Work	a) Jointly with the Contractor examine the Works and the Plant and Materials affected by the suspension. b) Make a written record of all making good required to be carried out by the Contractor.	a) After permission or instruction to proceed is given. b) None.	None.

THE OBLIGATIONS OF THE EMPLOYER'S REPRESENTATIVE (continued)

CLAUSE	OBLIGATIONS	TIME FRAME	SPECIFIC CONSEQUENCES OF NON-COMPLIANCE
10 Operation Service			
10.2 Commencement of Operation Service	In the case that the Contractor provides notification that he wishes to modify a previously submitted document, review the modification and if appropriate, give consent to proceed.	None.	None.
10.6 Delays and Interruptions during the Operation Service	a) Following a period of suspension by the Employer, jointly examine the Works with the Contractor. b) Make a written record of all making good required to be carried out by the Contractor.	a) After permission to proceed is given. b) None.	None.
11 Testing			
11.1 Testing of the Works	a) After receiving Notice from the Contractor, instruct the Contractor of the day or days that the Tests on Completion of Design-Build are to take place. b) In considering the results of the Tests on Completion of Design-Build, make due allowances for the effect of any use of the Works by the Employer.	a) So that the Tests may be carried out within 14 days of the date that the Contractor has notified that he will be ready to carry out the Tests. b) None.	None.
11.5 Completion of the Works and Sections	Issue the Commissioning Certificate to the Contractor, or reject the application for a Commissioning Certificate, giving reasons.	Within 28 days of receiving the Contractor's application.	The Commissioning Certificate shall be deemed to have been issued.
11.6 Commissioning of Parts of the Works	In the case of a Commissioning Certificate being issued for part of the Works, agree or determine reduced delay damages for the remainder of the Works and/or Section.	None.	None.

CLAUSE	OBLIGATIONS	TIME FRAME	SPECIFIC CONSEQUENCES OF NON-COMPLIANCE
11.7 Commissioning Certificate	Subject to the provisions of Sub-Clause 11.5 (*Completion of the Works and Sections*), issue the Commissioning Certificate.	Within 28 days of the Contractor's application.	The Commissioning Certificate shall be deemed to have been issued (Sub-Clause 11.5).
11.8 Joint Inspection Prior to Contract Completion	a) Jointly with the Contractor, carry out an inspection of the Works. b) Instruct the Contractor to carry out the Tests Prior to Contract Completion.	a) Not less than two years prior to the expiry date of the Operation Service Period. b) Upon satisfactory completion of maintenance works identified on the inspection report.	None.
11.9 Procedure for Tests Prior to Contract Completion	a) After receiving Notice from the Contractor, instruct the Contractor of the day or days that the Tests Prior to Contract Completion are to take place. b) Jointly with the Contractor, compile and evaluate the results of the Tests. c) Notify the Employer and the Contractor that the Works are complete and ready for final inspection.	a) So that the Tests may be commenced within 14-days of the date that the Contractor has notified that he will be ready to carry out the Tests. b) None. c) As soon as the Employer's Representative is satisfied that the Contractor has satisfied the requirements of the Tests.	None.
11.10 Delayed Tests Prior to Contract Completion	If the Contractor fails to commence the Tests, give Notice to the Contractor that others may be instructed to undertake the Tests.	None.	None.

THE OBLIGATIONS OF THE EMPLOYER'S REPRESENTATIVE (continued)

CLAUSE	OBLIGATIONS	TIME FRAME	SPECIFIC CONSEQUENCES OF NON-COMPLIANCE
12 Defects			
12.3 Failure to Remedy Defects	a) In the case that the Contractor fails to remedy any defect and the Employer's Representative fixes a date by which the defect should be remedied, give Notice to the Contractor. b) In the case that the Contractor fails to remedy any defect and if required by the Employer, agree or determine a reasonable reduction in the Contract Price.	a) Within a reasonable time. b) None.	None.
12.6 Contractor to Search	In the case that the Contractor has searched for a defect that is found not to be the responsibility of the Contractor, agree or determine the Cost Plus Profit of the search.	None.	None.
13 Variations and Adjustments			
13.1 Right to Vary	In the case that the Contractor gives Notice that the Contractor cannot readily obtain the Goods required for a Variation, cancel, confirm or vary the instruction.	None.	None.
13.3 Variation Procedure	a) Respond to the Contractor's Variation proposals with approval, disapproval or comments. b) Issue instructions to execute Variations, including any requirement for the recording of costs. c) Agree or determine adjustments to the Contract Price and Schedule of Payments for Variations.	a) As soon as practicable after receiving the proposal. b) None. c) None.	a) If the Variation is instructed during the Operation Service Period, the Contractor is not obliged to proceed with the Variation (Sub-Clause 13.1). b) None. c) None.

CLAUSE	OBLIGATIONS	TIME FRAME	SPECIFIC CONSEQUENCES OF NON-COMPLIANCE
13.5 Provisional Sums	Give instructions for the use of Provisional Sums.	None.	None.
13.6 Adjustments for Changes in Legislation	In the case of a Notice and claim for delay and/or additional Cost being received, respond to the claim and agree or determine the matters.	Within 42 days after receiving a claim or any further particulars (Sub-Clause 20.1).	Either Party may refer the matter to the DAB (Sub-Clause 20.1).
13.7 Adjustments for Changes in Technology	a) In the case of a Notice and claim for delay and/or additional Cost being received, respond to the claim and agree or determine the matters. b) Where appropriate, issue a Variation to the Contractor with details of the required changes.	a) Within 42 days after receiving a claim or any further particulars (Sub-Clause 20.1). b) None.	Either Party may refer the matter to the DAB (Sub-Clause 20.1).
14 Contract Price and Payment			
14.2 Advance Payment	Issue an Interim Payment Certificate for the advance payment.	After receiving an application under Sub-Clause 14.3 and after the Employer receives the Performance Security and an advance payment guarantee.	None.
14.3 Application for Advance and Interim Payment Certificates	Cooperate with the Contractor to agree and approve the form of the Statements.	None.	None.
14.6 Payment for Plant and Materials intended for the Works	Determine and certify an amount for Plant and Materials which have been sent to the Site for incorporation in the Permanent Works.	For inclusion in each Interim Payment Certificate.	None.

THE OBLIGATIONS OF THE EMPLOYER'S REPRESENTATIVE (continued)

CLAUSE	OBLIGATIONS	TIME FRAME	SPECIFIC CONSEQUENCES OF NON-COMPLIANCE
14.7 Issue of Advance and Interim Payment Certificates	a) Issue to the Employer an Interim Payment Certificate for the advance payment. b) Issue to the Employer an Interim Payment Certificate. c) In the case that the certified amount would be less than the minimum amount of Interim Payment Certificates stated in the Appendix to Tender, give Notice to the Contractor.	a) After the Employer has received and approved the Performance Security and within 14 days after receiving the application from the Contractor. b) Within 28 days after receiving an application from the Contractor. c) None.	If late certification results in the Employer not making payment within the stated period, the Contractor is entitled to receive financing charges (Sub-Clause 14.9).
14.10 Payment of Retention Money	Certify the first half of the Retention Money.	a) When the Commissioning Certificate has been issued.	If late certification results in the Employer not making payment within the stated period, the Contractor is entitled to receive financing charges (Sub-Clause 14.9).
14.11 Application for Final Payment Certificate Design-Build	a) In the case that the Employer's Representative disagrees with or cannot verify any part of the Final Statement Design-Build, attempt to agree such matters with the Contractor. b) Issue a Final Payment Certificate Design-Build for the agreed amount. c) In the case that the Parties cannot agree or if the Contractor has failed to submit his application for payment within the due time, issue an Interim Payment Certificate for the amount that the Employer's Representative considers due.	None.	None.

CLAUSE	OBLIGATIONS	TIME FRAME	SPECIFIC CONSEQUENCES OF NON-COMPLIANCE
14.12 Issue of Final Payment Certificate Design-Build	Issue, to the Employer the Final Payment Certificate Design-Build.	Within 28 days after receiving the Final Statement Design-Build and written undertaking in accordance with Sub-Clause 14.11.	If late certification results in the Employer not making payment within the stated period, the Contractor is entitled to receive financing charges (Sub-Clause 14.9).
14.15 Issue of Final Payment Certificate Operation Service	a) Issue, to the Employer the Final Payment Certificate Operation Service. b) In the case that the Employer's Representative disagrees with or cannot verify any part of the Final Statement Operation Service, attempt to agree such matters with the Contractor. c) Issue a Final Payment Certificate Operation Service for the agreed amount. d) In the case that the Parties cannot agree, issue a Final Payment Certificate Operation Service for the amount that the Employer's Representative considers due.	a) Within 28 days after receiving the Final Statement Operation Service and written discharge in accordance with Sub-Clause 14.14. b) None. c) None. d) None.	If late certification results in the Employer not making payment within the stated period, the Contractor is entitled to receive financing charges (Sub-Clause 14.9).
15 Termination by Employer			
15.1 Notice to Correct	In the case that the Contractor fails to carry out any obligation under the Contract, require the Contractor by Notice to make good the failure within the time specified within the Notice.	None.	None.
15.3 Valuation at Date of Termination for Contractor's Default	Agree or determine the value of the Works, Goods, Contractor's Documents and any other sums due to the Contractor for work executed in accordance with the Contract.	As soon as practicable after a Notice of termination.	None.

THE OBLIGATIONS OF THE EMPLOYER'S REPRESENTATIVE (continued)

CLAUSE	OBLIGATIONS	TIME FRAME	SPECIFIC CONSEQUENCES OF NON-COMPLIANCE
15.5 Valuation at Date of Termination for Employer's Convenience	Agree or determine the value of the Works, Goods, Contractor's Documents and any other sums due to the Contractor for work executed in accordance with the Contract.	As soon as practicable after a Notice of termination.	None.
16 Suspension and Termination by Contractor			
16.1 Contractor's Entitlement to Suspend Work	In the case of a Notice and claim for delay or cost being received, respond to the claim and agree or determine the matters.	Within 42 days after receiving a claim or any further particulars (Sub-Clause 20.1).	Either Party may refer the matter to the DAB (Sub-Clause 20.1).
17 Risk Allocation			
17.6 Consequences of the Employer's Risks of Damage	In the case of a Notice being received, determine the amounts due.	Within 42 days after receiving a claim or any further particulars (Sub-Clause 20.1).	Either Party may refer the matter to the DAB (Sub-Clause 20.1).
18 Exceptional Risks			
18.4 Consequences of an Exceptional Event	In the case of a Notice and claim for delay or cost being received, respond to the claim and agree or determine the matters.	Within 42 days after receiving a claim or any further particulars (Sub-Clause 20.1).	Either Party may refer the matter to the DAB (Sub-Clause 20.1).
18.5 Optional Termination, Payment and Release	In the case of termination, determine the value of the work done and issue a payment certificate.	Upon termination.	None.

CLAUSE	OBLIGATIONS	TIME FRAME	SPECIFIC CONSEQUENCES OF NON-COMPLIANCE
20 Claims, Disputes and Arbitration			
20.1 Contractor's Claims	a) In the case of a claim being received, agree or determine the extension of the Time for Completion and/or the additional payment (if any) with detailed comments. b) In the case of additional particulars being requested, respond on the contractual or other aspects of the claim. c) Include such amounts for any claim as have been reasonably substantiated as due within each Payment Certificate.	a) Within 42 days after receiving a claim or any further particulars. b) Within 42 days after receiving the claim or any further particulars. c) As appropriate.	The matter may be referred to the DAB.
20.2 Employer's Claims	a) In the case that the Employer considers himself entitled to any payment, give Notice and particulars to the Contractor (this action may also be undertaken by the Employer). b) Agree or determine the amount (if any) to which the Employer is entitled.	As soon as practicable after the Employer becomes aware or should have become aware of the event or circumstances giving rise to the claim.	
20.6 Obtaining Dispute Adjudication Board's Decision	Comply with decisions given by the DAB.	Promptly.	Failure to comply may be referred to arbitration (Sub-Clause 20.9)

GENERAL CONDITIONS OF DISPUTE ADJUDICATION AGREEMENT

Procedural Rules for Dispute Adjudication Board Members			
3.	Attend Site visits by the DAB.	None.	None.

THE OBLIGATIONS OF THE EMPLOYER'S REPRESENTATIVE

Gold Book

THE OBLIGATIONS OF THE DISPUTE ADJUDICATION BOARD

CLAUSE	OBLIGATIONS	TIME FRAME	SPECIFIC CONSEQUENCES OF NON-COMPLIANCE
GENERAL CONDITIONS			
20 Claims, Disputes and Arbitration			
20.6 Obtaining Dispute Adjudication Board's Decision	Give a reasoned decision on any Dispute referred to the DAB.	Within 84 days after receiving the other Party's response to the referral of a Dispute.	Either Party may give a Notice of dissatisfaction and commence arbitration.
GENERAL CONDITIONS OF DISPUTE ADJUDICATION AGREEMENT			
3 Warranties	a) Be impartial and independent of the Employer and the Contractor and the Employer's Representative. b) Disclose to the Parties and to the Other Members, any fact or circumstance that might appear inconsistent with his/her warranty and agreement of impartiality and independence.	a) None. b) Promptly.	None.

CLAUSE	OBLIGATIONS	TIME FRAME	SPECIFIC CONSEQUENCES OF NON-COMPLIANCE
4 General Obligations of the Member	a) Have no interest, financial or otherwise in the Parties, nor any financial interest in the Contract. b) Not previously have been employed as a consultant or otherwise by the Parties, except as disclosed in writing. c) Disclose in writing to the Parties and the Other Members, any professional or personal relationships with any director, officer or employee of the Parties and any previous involvement in the overall project of which the Contract forms part. d) Not, for the duration of the Dispute Adjudication Agreement, to be employed as a consultant or otherwise by the Parties, except as may be agreed in writing. e) Comply with the Procedural Rules and with Sub-Clause 20.5 of the Conditions of Contract. f) Not to give advice to the Parties, the Employer's Personnel or the Contractor's Personnel concerning the conduct of the Contract, other than in accordance with the Procedural Rules. g) Not to enter into discussions or make any agreement with the Employer or the Contractor regarding employment by any of them after ceasing to act under the Dispute Adjudication Agreement. h) Ensure his/her availability for all Site visits and hearings as are necessary. i) Become conversant with the Contract and with the progress of the Works and maintain documents received in a current working file. j) Treat the details of the Contract and all the DAB's activities and hearings as private and confidential. k) Be available to give advice and opinions on any matter related to the Contract when requested by both the Employer and the Contractor.	None.	None.
6 Payment	a) Submit invoices for payment of the monthly retainer and air fares. b) Submit invoices for other expenses and for daily fees.	a) Quarterly in advance. b) Following the conclusion of a Site visit or hearing.	None.

THE OBLIGATIONS OF THE DISPUTE ADJUDICATION BOARD

Gold Book

THE OBLIGATIONS OF THE DISPUTE ADJUDICATION BOARD (continued)

CLAUSE	OBLIGATIONS	TIME FRAME	SPECIFIC CONSEQUENCES OF NON-COMPLIANCE
Annex – Procedural Rules for Dispute Adjudication Board Members			
1.	Visit the Site.	At intervals of not more than 140 days including time of critical construction events and not less than 70 days.	
2.	Agree the timing and agenda for each Site visit with the Employer and the Contractor.	None.	
3.	Prepare a report of the DAB activities during the Site visit and send copies to the Employer and the Contractor.	At the conclusion of each Site visit.	
4.	Copy all communications between the DAB and the Employer or the Contractor to the other Party.	None.	None.
5.	If a Dispute is referred to the DAB, proceed in accordance with Sub-Clauses 20.6 and 20.1.	None.	None.
5(a).	a) Act fairly and impartially as between the Parties. b) Give each of the Parties a reasonable opportunity of putting his case and responding to the other's case.	None.	None.
5(b).	Adopt procedures suitable to the Dispute, avoiding unnecessary delay or expense.	None.	None.

CLAUSE	OBLIGATIONS	TIME FRAME	SPECIFIC CONSEQUENCES OF NON-COMPLIANCE
7.	In the case that the DAB find and agree that a decision contained errors, advise the Employer and the Contractor of the error and issue an addendum to the decision.	Within 14 days of giving the decision.	None.
8.	In the case that either Party believes that a decision contains an ambiguity and requests clarification, issue a response.	Within 14 days of receiving the request.	None.
11.	a) Not express any opinions during any hearing concerning the merits of any arguments advanced by the Parties. b) Make and give a decision in accordance with Sub-Clause 20.6, or as otherwise agreed by the Employer and the Contractor in writing.	None.	None.
12.	If the DAB comprises three persons: I. Convene in private after a hearing. II. Endeavour to reach a unanimous decision.	None.	None.

THE OBLIGATIONS OF THE DISPUTE ADJUDICATION BOARD

Gold Book

Chapter 7
The Green Book
Short form of Contract, First Edition 1999

The FIDIC Contracts: Obligations of the Parties, First Edition. Andy Hewitt.
© 2014 John Wiley & Sons, Ltd. Published 2014 by John Wiley & Sons, Ltd.

THE OBLIGATIONS OF THE EMPLOYER

CLAUSE	OBLIGATIONS	TIME FRAME	SPECIFIC CONSEQUENCES OF NON-COMPLIANCE
GENERAL CONDITIONS			
1 General Provisions			
1.5 Communications	a) Issue written communications in the language stated in the Appendix. b) Not to unreasonably withhold any notice, instruction or other communication.	None.	None.
2 The Employer			
2.1 Provision of Site	Provide the Site and right of access thereto.	At the time stated in the Appendix.	None.
2.2 Permits and Licences	If requested by the Contractor, assist him in applying for permits, licenses or approvals.	None.	None.
3 Employer's Representatives			
3.1 Authorised Person	If the authorised person is not stated in the Appendix, notify the Contractor of the person who has the authority to act for the Employer.	None.	None.
3.2 Employer's Representative	a) In the case that the Employer appoints a firm or individual to carry out certain duties and the firm or individual is not named in the Appendix, notify the Contractor. b) Notify the Contractor of the delegated duties and authority of the appointed firm or individual.	None.	None.

Green Book

THE OBLIGATIONS OF THE EMPLOYER (continued)

CLAUSE	OBLIGATIONS	TIME FRAME	SPECIFIC CONSEQUENCES OF NON-COMPLIANCE
5 Design by Contractor			
5.1 Contractor's Design	In the case that the Contractor submits designs, notify the Contractor of any comments or, if the design is not in accordance with the Contract, reject the design stating the reasons.	Within 14 days of receipt of the design.	None.
7 Time for Completion			
7.3 Extension of Time	On receipt of an application for an extension to the Time for Completion, consider all supporting details and extend the Time for Completion as appropriate.	On receipt of the application.	None.
8 Taking Over			
8.2 Taking-Over Notice	a) Notify the Contractor when the Employer considers that the Contractor has completed the Works, stating the date accordingly. b) Take over the Works.	a) None. b) Upon the issue of the notice	None.
10 Variations and Claims			
10.3 Early Warning	Notify the Contractor of any circumstance which may delay or disrupt the Works, or which may give rise to a claim for additional payment.	As soon as becoming aware of the circumstance.	None.
10.5 Variation and Claim Procedure	a) Check and agree the value of the Contractor's itemised make-up of the value of Variations and claims. b) In the absence of agreement, determine the value.	None.	None.

CLAUSE	OBLIGATIONS	TIME FRAME	SPECIFIC CONSEQUENCES OF NON-COMPLIANCE
11 Contract Price and Payments			
11.3 Interim Payments	Pay the Contractor the amount shown in the Contractor's statement, less any retention and any amount for which the Employer has specified his reasons for disagreement.	Within 28 days of delivery of each statement.	Contractor shall be entitled to interest (Sub-Clause 11.8).
11.4 Payment of First Half of Retention	Pay the Contractor one half of the retention.	Within 14 days after issuing the Taking-Over Notice.	Contractor shall be entitled to interest (Sub-Clause 11.8).
11.5 Payment of Second Half of Retention	Pay the Contractor the remainder of the retention.	Within 14 days of the expiry of the period stated in the Appendix, or the remedying of notified defects, or completion of outstanding work, whichever is the later.	Contractor shall be entitled to interest (Sub-Clause 11.8).
11.6 Final Payment	a) Pay the Contractor any amount due under the final account. b) In the case that the Employer disagrees with any part of the Contractor's final account, specify the reasons for disagreement.	a) Within 42 days of the latest of the events listed in Sub-Clause 11.5. b) When making payment.	Contractor shall be entitled to interest (Sub-Clause 11.8).
11.7 Currency	Make payments to the Contractor in the currencies stated in the Appendix.	None.	None.
13 Risk and Responsibility			
13.2 Force Majeure	In the case of being prevented from performing the Employer's obligations by Force Majeure, notify the Contractor.	Immediately.	None.

THE OBLIGATIONS OF THE EMPLOYER

Green Book

THE OBLIGATIONS OF THE EMPLOYER (continued)

CLAUSE	OBLIGATIONS	TIME FRAME	SPECIFIC CONSEQUENCES OF NON-COMPLIANCE
15 Resolution of Disputes			
15.1 Adjudication	In the case of being unable to settle any dispute or difference amicably, refer the matter to adjudication in accordance with the Rules for Adjudication.	None.	None.
RULES FOR ADJUDICATION			
Appointment of Adjudicator	Jointly with the Contractor, ensure the appointment of the Adjudicator.	Within 14 days of the reference of a dispute.	The Contractor may apply for appointment by the appointing body named in the Contract.
Terms of Appointment	a) Not to call the Adjudicator as a witness to give evidence concerning any dispute. b) In the event of the Adjudicator not being able to carry out the duties agree a replacement Adjudicator jointly with the Contractor.	a) None. b) Within 14 days.	a) None. b) The Contractor may apply for appointment by the appointing body named in the Contract.
Payment	Reimburse the Contractor for half of the Adjudicator's fees.	None.	None.
Procedure for Obtaining Adjudicator's Decision	a) In the event that a dispute is referred, identify the dispute and refer to the Rules for Adjudication. b) Provide the Adjudicator with copies of any documentation and information that he may request. c) Communicate with the Adjudicator in the language of the Adjudicator's Agreement. d) Copy all communications to the Adjudicator to the Contractor.	a) None. b) Promptly. c) None. d) None.	None.

THE OBLIGATIONS OF THE CONTRACTOR

CLAUSE	OBLIGATIONS	TIME FRAME	SPECIFIC CONSEQUENCES OF NON-COMPLIANCE
GENERAL CONDITIONS			
1 General Provisions			
1.5 Communications	Issue written communications in the language stated in the Appendix.	None.	None.
1.6 Statutory Obligations	a) Comply with the laws of the countries where the activities are performed. b) Give all notices and pay all fees and other charges in respect of the Works.	None.	None.
2 The Employer			
2.3 Employer's Instructions	Comply with all instructions given by the Employer in respect of the Works.	None.	None.
4 The Contractor			
4.1 General Obligations	a) Carry out the Works properly and in accordance with the Contract. b) Provide all supervision, labour, Materials, Plant and Contractor's Equipment which may be required.	None.	None.
4.2 Contractor's Representative	Submit to the Employer for consent, the name and particulars of the person authorised to receive instructions.	None.	None.
4.3 Subcontracting	a) Not to subcontract the whole of the Works. b) Not to subcontract any part of the Works without the consent of the Employer.	None.	None.

THE OBLIGATIONS OF THE CONTRACTOR (continued)

CLAUSE	OBLIGATIONS	TIME FRAME	SPECIFIC CONSEQUENCES OF NON-COMPLIANCE
4.4 Performance Security	If stated in the Appendix, deliver to the Employer a performance security in a form and from a third party approved by the Employer.	Within 14 days of the Commencement Date.	The Employer may withhold payment (Sub-Clause 11.3).
5 Design by Contractor			
5.1 Contractor's Design	a) Carry out design to the extent specified. b) Submit to the Employer all designs prepared. c) Not to construct any element of the permanent work designed by the Contractor within 14 days after the design has been submitted to the Employer, or where the design has been rejected. d) Amend and resubmit any design that has been rejected. e) Resubmit all designs that have been commented on, taking the comments into consideration.	a) None. b) Promptly. c) Within 14 days after the design has been submitted to the Employer. d) Promptly. e) None.	None.
7 Time for Completion			
7.1 Execution of the Works	a) Commence the Works. b) Proceed expeditiously and without delay. c) Complete the Works.	a) On the Commencement Date. b) None. c) Within the Time for Completion	a) None. b) None. c) The Contractor shall pay the Employer the amount stated in the Appendix for each day which he fails to complete the Works (Sub-Clause 7.4).
7.2 Programme	Submit to the Employer a programme for the Works in the form stated in the Appendix.	Within the time stated in the Appendix.	None.

CLAUSE	OBLIGATIONS	TIME FRAME	SPECIFIC CONSEQUENCES OF NON-COMPLIANCE
8 Taking Over			
8.2 Taking-Over Notice	Complete any outstanding work and clear the Site.	Promptly upon the Employer taking over the Works.	None.
9 Remedying Defects			
9.1 Remedying Defects	Remedy any defects.	Upon notice from the Employer.	The Employer is entitled to carry out all necessary work at the Contractor's cost.
10 Variations and Claims			
10.2 Valuation of Variations	If the Employer instructs Variations to be valued at daywork rates, keep records of hours of labour, Contractor's Equipment and Materials used.	None.	None.
10.3 Early Warning	a) Notify the Employer of any circumstance which may delay or disrupt the Works, or which may give rise to a claim for additional payment. b) Take all reasonable steps to minimise these effects.	a) As soon as becoming aware of the circumstance. b) None.	None.
10.5 Variation and Claim Procedure	Submit an itemised make-up of the value of Variations and claims.	Within 28 days of the instruction or of the event giving rise to the claim.	None.
11 Contract Price and Payment			
11.2 Monthly Statements	Submit a statement showing the amounts to which the Contractor considers himself entitled.	Each month.	The Employer is not obliged to pay the Contractor (Sub-Clause 11.3).
11.6 Final Payment	Submit a final account together with any documentation reasonably required to enable the Employer to ascertain the final contract value.	Within 42 days of the latest of the events listed in Sub-Clause 11.5.	The Employer is not obliged to pay the final amount due.

THE OBLIGATIONS OF THE CONTRACTOR

Green Book

THE OBLIGATIONS OF THE CONTRACTOR (continued)

CLAUSE	OBLIGATIONS	TIME FRAME	SPECIFIC CONSEQUENCES OF NON-COMPLIANCE
12 Default			
12.1 Default by Contractor	In the case that the Employer gives a notice of termination in accordance with this clause, demobilise from the Site leaving behind Materials and Plant and any Contractor's Equipment which the Employer instructs is to be used until the completion of the Works.	None.	None.
12.2 Default by Employer	In the case that the Contractor terminates the Contract in accordance with this clause, demobilise from the Site.	None.	None.
12.3 Insolvency	a) In the case that either Party terminates the Contract because of insolvency, demobilise from the Site. b) In the case of termination as a result of the Contractor's insolvency, leave behind any Contractor's Equipment which the Employer instructs is to be used until the completion of the Works.	None.	None.
13 Risk and Responsibility			
13.1 Contractor's Care of the Works	a) Take full responsibility for the care of the Works. b) In the case of any loss or damage, rectify such loss or damage so that the Works conform with the Contract. c) Indemnify the Employer, the Employer's contractor's, agents and employees against all loss or damage happening to the Works and against all claims or expense arising out of the Works caused by a breach of the Contract, by negligence, or any other default of the Contractor, his agents or employees.	From the Commencement Date until the date of the Employer's Taking-Over Notice.	None.

CLAUSE	OBLIGATIONS	TIME FRAME	SPECIFIC CONSEQUENCES OF NON-COMPLIANCE
13.2 Force Majeure	a) In the case of being prevented from performing the Contractor's obligations by Force Majeure, notify the Employer. b) If necessary, suspend the execution of the Works. c) To the extent agreed with the Employer, demobilise the Contractor's Equipment.	a) Immediately. b) None. c) None.	None.
14 Insurance			
14.1 Extent of Cover	Effect and thereafter maintain the insurances specified in this sub-clause in the joint names of the Parties.	Prior to commencing the Works.	The Employer may effect the insurance and recover the premiums from monies due to the Contractor.
14.2 Arrangements	a) Ensure that the insurances are issued by insurers and in terms approved by the Employer. b) Provide the Employer with evidence that any required policy is in force and that the premiums have been paid.	None.	The Employer may effect the insurance and recover the premiums from monies due to the Contractor.
15 Resolution of Disputes			
15.1 Adjudication	In the case of being unable to settle any dispute or difference amicably, refer the matter to adjudication in accordance with the Rules for Adjudication.		

THE OBLIGATIONS OF THE CONTRACTOR

Green Book

THE OBLIGATIONS OF THE CONTRACTOR (continued)

CLAUSE	OBLIGATIONS	TIME FRAME	SPECIFIC CONSEQUENCES OF NON-COMPLIANCE
RULES FOR ADJUDICATION			
Appointment of Adjudicator	Jointly with the Employer, ensure the appointment of the Adjudicator.	Within 14 days of the reference of a dispute.	The Employer may apply for appointment by the appointing body named in the Contract.
Terms of Appointment	a) Not to call the Adjudicator as a witness to give evidence concerning any dispute. b) In the event of the Adjudicator not being able to carry out the duties, agree a replacement Adjudicator jointly with the Contractor.	a) None. b) Within 14 days.	a) None. b) The Contractor may apply for appointment by the appointing body named in the Contract.
Payment	Pay the Adjudicator's fees.	Within 28 days of receipt of the Adjudicator's invoices.	a) The Adjudicator may suspend work. b) The Employer may pay the Adjudicator and recover the sum from the Contractor.
Procedure for Obtaining Adjudicator's Decision	a) In the event that a dispute is referred, identify the dispute and refer to the Rules for Adjudication. b) Provide the Adjudicator with copies of any documentation and information that he may request. c) Communicate with the Adjudicator in the language of the Adjudicator's Agreement. d) Copy all communications to the Adjudicator to the Employer.	a) None. b) Promptly. c) None. d) None.	None.

CLAUSE	OBLIGATIONS	TIME FRAME	SPECIFIC CONSEQUENCES OF NON-COMPLIANCE
RULES FOR ADJUDICATION			
Terms of Appointment	a) Be impartial and independent of the Parties. b) Disclose in writing anything that could affect impartiality or independence. c) Not give advice to the Parties or their representatives, other than in accordance with the Rules for Adjudication. d) Treat all details as confidential and do not disclose the same without prior written consent of the Parties. e) Not assign or delegate work under the Rules for Adjudication, or engage legal or technical assistance. f) In the case of resignation, give 28 days' notice to the Parties. g) In the case of breach of impartiality or an act of bad faith resulting in proceedings or decisions being rendered void, reimburse each of the Parties for fees and expenses paid	a) None. b) Immediately. c) None. d) None. e) None. f) None. g) None.	None.
Payment	a) Submit invoices for monthly retainer fees if applicable. b) Submit invoices for daily fees and expenses.	a) Quarterly in advance. b) Following the conclusion of a Site visit or hearing.	None.
Procedure for Obtaining Adjudicator's Decision	a) Act as an impartial expert, not as an arbitrator. b) Copy all communications to both Parties. c) Give written notice of decisions, including reasons, to the Parties	a) None. b) None. c) Within 56 days after receiving the reference.	None.

Chapter 8
The White Book
Client/Consultant Model Services Agreement, Fourth Edition 2006

The FIDIC Contracts: Obligations of the Parties, First Edition. Andy Hewitt.
© 2014 John Wiley & Sons, Ltd. Published 2014 by John Wiley & Sons, Ltd.

THE OBLIGATIONS OF THE CLIENT

CLAUSE	OBLIGATIONS	TIME FRAME	SPECIFIC CONSEQUENCES OF NON-COMPLIANCE
GENERAL CONDITIONS			
1 General Provisions			
1.3 Communications	a) Issue written communications in the language stated in the Particular Conditions. b) Not to unreasonably withhold any notice, instruction or other communication.	None.	None.
1.6 Assignments and Sub-Contracts	Not to assign obligations under the Agreement without the written consent of the Consultant.	None.	None.
2 The Client			
2.1 Information	Give to the Consultant free of cost all information which may pertain to the Services which the Client is able to obtain.	Within a reasonable time.	None.
2.2 Decisions	Give decisions in writing on all matters referred in writing by the Consultant.	Within a reasonable time so as not to delay the Services.	None.
2.3 Assistance	Do all in the Client's power, to assist in the following, in connection with the Services: a) provision of documents for entry, residency, working and exit; b) provision of unobstructed access; c) import, export and customs clearance; d) repatriation in the case of emergencies; e) provision of authority necessary for the import of foreign currency and the export of money earned; f) provision of access to organisations for collection of information.	None.	None.

White book

THE OBLIGATIONS OF THE CLIENT (continued)

CLAUSE	OBLIGATIONS	TIME FRAME	SPECIFIC CONSEQUENCES OF NON-COMPLIANCE
2.4 Client's Financial Arrangements	a) Submit reasonable evidence that financial arrangements have been made and are being maintained which will enable the Client to pay the Consultant's fees. b) If it is intended to make any material change to the financial arrangements, give notice to the Consultant with detailed particulars	a) Within 28 days of receiving a request from the Consultant. b) None.	None.
2.5 Equipment and Facilities	Make available, free of cost, the equipment and facilities described in Appendix 2.	None.	None.
2.6 Supply of Client's Personnel	Arrange for the selection and provision of personnel in accordance with Appendix 2.	None.	None.
2.7 Client's Representative	Designate an official or individual as the Client's representative for the administration of the Agreement.	None.	None.
2.8 Services of Others	Arrange for the provisions of services from others in accordance with Appendix 2.	None.	None.
2.9 Payment for Services	Pay the Consultant for the Services in accordance with Section 5 of the Agreement.	None.	The Consultant may terminate the Agreement or suspend the performance of the Services (Sub-Clause 4.6.3).
3 The Consultant			
3.5 Supply of Personnel	Not to unreasonably withhold the acceptance of the Consultant's Personnel who are proposed to work in the Country.	None.	None.

CLAUSE	OBLIGATIONS	TIME FRAME	SPECIFIC CONSEQUENCES OF NON-COMPLIANCE
3.7 Changes in Personnel	a) Request any change of personnel provided by the Consultant in writing, stating the reasons for the request. b) Bear the cost of the replacement unless it is agreed that misconduct or inability to perform satisfactorily is accepted.	None.	None.
4 Commencement, Completion, Variation and Termination			
4.4 Delays	In the case that the Services are impeded or delayed by the Client or his contractors so as to increase the scope, cost or duration of the Services, increase the time of completion accordingly.	None.	None.
4.6 Abandonment, Suspension or Termination	In the case of suspending all or part of the Services or terminating the Agreement, give notice to the Consultant.	56 days prior to the time when the suspension or termination is to take effect.	None.
5 Payment			
5.1 Payment to the Consultant	a) Pay the Consultant for Normal Services in accordance with the Conditions and with the details stated in Appendix 3. b) Pay the Consultant for Additional Services at rates and prices which are given or based on those in Appendix 3 or otherwise agreed. c) Pay the Consultant in respect of Exceptional Services.	Within 28 days of the Consultant's invoice unless otherwise stated in the Particular Conditions (Sub-Clause 5.2)	1. The Consultant may terminate the Agreement or suspend the performance of the Services (Sub-Clause 4.6.3). 2. The Consultant shall be paid Agreed Compensation at the rate stated in the Particular Conditions (Sub-Clause 5.2.2).
5.2 Time for Payment	Not to withhold payment of any fee properly due without giving to the Consultant a notice of intention to withhold payment with reasons.	Notice to be given no later than 4 days prior to the date on which the payment becomes due.	The Consultant shall have an enforceable right to such payment.

THE OBLIGATIONS OF THE CLIENT

White book

THE OBLIGATIONS OF THE CLIENT (continued)

CLAUSE	OBLIGATIONS	TIME FRAME	SPECIFIC CONSEQUENCES OF NON-COMPLIANCE
5.4 Third Party Charges on the Consultant	a) Whenever possible, arrange that exemption is granted to the Consultant and his personnel who are not normally resident in the Country, from any payment required by the Government or other authorised parties in the Country, in respect of the items listed in the sub-clause. b) In the case that the Client is unsuccessful in arranging exemption, reimburse the Consultant for such payments properly made	None.	None.
5.5 Disputed Invoices	a) In the case that any item or part of an invoice is contested, give notice of intention to withhold payment with reasons. b) Not to delay payment on the remainder of the invoice.	None.	a) None. b) The Consultant shall be paid Agreed Compensation at the rate stated in the Particular Conditions (Sub-Clause 5.2.2).
6 Liabilities			
6.3 Limit of Compensation	In the case that the Client makes a claim for compensation against the Consultant that is not established, reimburse the Consultant for his costs incurred as a result of the claim.	None.	None.
6.4 Indemnity	Indemnify the Consultant against the adverse effects of all claims including claims by third parties which arise out of or in connection with the Agreement, except so far as they are covered by the insurances arranged under Clause 7.1.	None.	None.

CLAUSE	OBLIGATIONS	TIME FRAME	SPECIFIC CONSEQUENCES OF NON-COMPLIANCE
8 Disputes and Arbitration			
8.1 Amicable Dispute Resolution	a) In the case of a dispute in connection with the Agreement, ensure that representatives of the Client with authority to settle the dispute, meet in good faith to attempt to resolve the dispute. b) In the case that the dispute is not resolved, attempt to settle it by means of mediation.	a) Within 14 days of a written request by the Consultant. b) None.	None.
8.2 Mediation	a) Attempt to agree upon a neutral mediator from a list held by the independent mediation centre named in the Particular Conditions. b) Not to commence arbitration of any dispute until any attempt has been made to settle the dispute by mediation and the mediation has terminated or the Consultant has failed to participate in the mediation.	a) Within 14 days of a notice from the Consultant requesting mediation. b) None	a) The Consultant may request that the President of FIDIC appoint a mediator. b) None.

THE OBLIGATIONS OF THE CLIENT

White book

THE OBLIGATIONS OF THE CONSULTANT

CLAUSE	OBLIGATIONS	TIME FRAME	SPECIFIC CONSEQUENCES OF NON-COMPLIANCE
GENERAL CONDITIONS			
1 General Provisions			
1.3 Communications	a) Issue written communications in the language stated in the Particular Conditions. b) Not to unreasonably withhold any notice or other communication.	None.	None.
1.6 Assignments and Sub-Contracts	a) Not to assign the benefits, other than money, from the Agreement without the written consent of the Client. b) Not to assign obligations under the Agreement without the written consent of the Client. c) Not to initiate or terminate any sub-contract for performance of all or part of the Services without the written consent of the Client.	None.	None.
1.10 Corruption and Fraud	a) Comply with all applicable laws, rules, regulations and orders of any applicable jurisdiction, including the OECD Convention on Combatting Bribery of Foreign Public Officials in International Business Transactions. b) Notify the Client in the event that the Consultant receives a request from any public official requesting illicit payments.	a) None. b) Immediately.	None.
2 The Client			
2.6 Supply of Client's Personnel	In the case that the Client cannot supply personnel for which he is responsible and it is agreed to be necessary, arrange for such supply as an Additional Service.	None.	None.
2.8 Services of Others	Co-operate with suppliers of other services.	None.	None.

CLAUSE	OBLIGATIONS	TIME FRAME	SPECIFIC CONSEQUENCES OF NON-COMPLIANCE
3 The Consultant			
3.1 Scope of Services	Perform the Services as stated in Appendix 1.	In accordance with the Commencement Date and the Time for Completion (Sub-Clause 4.2.1).	None.
3.3 Duty of Care and Exercise of Authority	Exercise reasonable skill, care and diligence in the performance of the Consultant's obligations under the Agreement.	None.	None.
3.6 Representatives	a) Designate an official or individual to be the Consultant's representative for the administration of the Agreement. b) If required by the Client, designate an individual to liaise with the Client's representative in the Country.	None.	None.
3.7 Changes in Personnel	Arrange for the replacement of any of the personnel by a person of comparable competence.	As soon as reasonably possible.	None.
4 Commencement, Completion, Variation and Termination			
4.2 Commencement and Completion	a) Commence the Services on the Commencement Date. b) Proceed with the Services in accordance with the Time Schedule in Appendix 4. c) Complete the Services within the Time for Completion.	None.	None.
4.3 Variations	If requested by the Client, submit proposals for varying the Services.	None.	None.
4.4 Delays	In the case that the Services are impeded or delayed by the Client or his contractors so as to increase the scope, cost or duration of the Services, inform the Client of the circumstances and probable effects.	None.	None.
4.5 Changed Circumstances	If circumstances arise which make it irresponsible or impossible for the Consultant to perform the Services, give notice to the Client.	Promptly.	None.

THE OBLIGATIONS OF THE CONSULTANT

White book

THE OBLIGATIONS OF THE CONSULTANT (*continued*)

CLAUSE	OBLIGATIONS	TIME FRAME	SPECIFIC CONSEQUENCES OF NON-COMPLIANCE
4.6 Abandonment, Suspension or Termination	In the case of receiving notice to suspend all or part of the Services or to terminate the Agreement, make arrangements to stop the Services and minimise expenditure.	Immediately.	None.
5 Payment			
5.4 Third Party Charges on the Consultant	a) Not to dispose of goods imported for the Services in the Country without the Client's approval. b) Not to export goods imported for the Services without payment to the Client of any refund or rebate recoverable and received from the Government or Authorised third parties	None.	None.
5.6 Independent Audit	Maintain up-to-date records which clearly identify relevant time and expense and make these available to the Client on reasonable request.	None.	None.
6 Liabilities			
6.3 Limit of Compensation	In the case that the Consultant makes a claim for compensation against the Client that is not established, reimburse the Client for his costs incurred as a result of the claim.	None.	None.
7 Insurance			
7.1 Insurance for Liability and Indemnity	Make reasonable efforts to insure or increase insurance against the listed liabilities.	None.	None.

CLAUSE	OBLIGATIONS	TIME FRAME	SPECIFIC CONSEQUENCES OF NON-COMPLIANCE
7.2 Insurance of Client's Property	Make reasonable efforts to insure on terms acceptable to the Client, against loss or damage to the property of the Client and against liabilities arising out of the use of such property.	None.	None.
8 Disputes and Arbitration			
8.1 Amicable Dispute Resolution	a) In the case of a dispute in connection with the Agreement, ensure that representatives of the Consultant with authority to settle the dispute meet in a good faith to attempt to resolve the dispute. b) In the case that the dispute is not resolved, attempt to settle it by means of mediation.	a) Within 14 days of a written request by the Client. b) None.	None.
8.2 Mediation	a) Attempt to agree upon a neutral mediator from a list held by the independent mediation centre named in the Particular Conditions. b) Not to commence arbitration of any dispute until any attempt has been made to settle the dispute by mediation and the mediation has terminated, or the Client has failed to participate in the mediation.	a) Within 14 days of a notice from the Client requesting mediation. b) None	a) The Client may request that the President of FIDIC appoint a mediator. b) None.

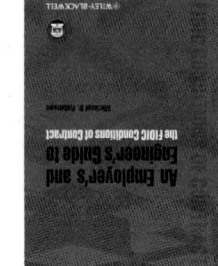

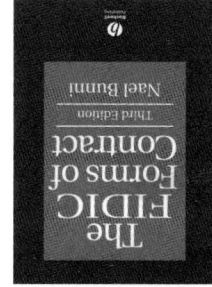